Sir William Roberts

On the digestive ferments and the preparation and use of

artificially digested food

Sir William Roberts

On the digestive ferments and the preparation and use of artificially digested food

ISBN/EAN: 9783337201227

Printed in Europe, USA, Canada, Australia, Japan

Cover: Foto ©berggeist007 / pixelio.de

More available books at **www.hansebooks.com**

ON THE

DIGESTIVE FERMENTS

AND THE

PREPARATION AND USE

OF

ARTIFICIALLY DIGESTED FOOD ;

BEING THE LUMLEIAN LECTURES FOR THE YEAR
1880.

Delivered before the Royal College of Physicians.

BY

WM. ROBERTS, M.D., F.R.S.,

FELLOW OF THE COLLEGE;
PHYSICIAN TO THE MANCHESTER ROYAL INFIRMARY; PROFESSOR OF CLINICAL
MEDICINE TO THE OWENS COLLEGE.

SECOND EDITION—REVISED—WITH ADDITIONS.

LONDON :
SMITH, ELDER, & CO., WATERLOO PLACE.
MANCHESTER:
J. E. CORNISH, 33 PICCADILLY.
1881.

PREFACE TO THE SECOND EDITION.

————

THE principal additions made to this edition are enclosed in square brackets, thus [ ]. One of these, on p. 12, has reference to the albuminoid matters contained in the liquid preparations of the proteolytic ferments; another, on p. 13, contains suggestions with regard to nomenclature. The directions for the preparation of articles of peptonised food have been extended and, at the same time, simplified—and a new method is added for the preparation of peptonised milk at the ordinary temperature of the sick room, which I think will prove acceptable to practitioners. In a foot-note at p. 59 I have given a brief sketch of a method of estimating quantitatively the dua activities of pancreatic extracts.

MANCHESTER, *June*, 1881.

CONTENTS.

LECTURE I.

LECTURE II.

iv.

LECTURE III.

INDEX.

LECTURE I.

SUMMARY.—Digestion is a Faculty or Function Common to Animals and Plants—Exterior Digestion—Interstitial Digestion—General Characters and Properties of the Digestive Ferments—Preparation of Artificial Digestive Juices—Diastasic Ferments and the Digestion of Starch—Theory of the Process—The Respective Shares of Saliva and Pancreatic Juice in the Digestion of Starch —When can Starch be said to be Fully Digested?—Absolute Energy of Diastase—Digestion of Cane-sugar—and the Inversive Ferment.

Digestion has been usually regarded as the special attribute of animals. They receive into their alimentary canal the food which they require for their sustenance in a crude form. It is there subjected to the action of certain ferments which transmute its elements, by a peculiar chemical process, into new forms which are fitted for absorption. Looked at in this restricted sense, plants have no digestive function. They possess no alimentary canal nor any vestige of a digestive apparatus. But when the matter is examined more profoundly it is seen that plants digest as well as animals, and that the process in both kingdoms of nature is fundamentally the same.

In order to understand this generalization—which was first propounded by Claude Bernard, and constitutes one of the most important fruits of his splendid labours*—it is necessary to recognize digestion under two types or conditions—namely, a digestion which takes place exteriorly at the surface of the organism, and a digestion which takes place interstitially in the interior of the organs and tissues.

Exterior digestion is that common process with which we are familar as taking place in the alimentary canal of animals, by which the crude food introduced from without is prepared for absorption.

Interstitial digestion, on the other hand, is that more recondite process by which the reserves of food lodged in the interior of plants and animals are modified and made available for the purposes of nutrition.

These two types of digestion are essentially alike both as regards the agents and the processes by which they are carried

---

* Claude Bernard—*Leçons sur les phénomènes de la vie*, T. II. Edited after his death by Dastre—Paris, 1879.

out—and although one type of digestion is more developed in
the animal kingdom and the other type more developed in the
vegetable kingdom, both types are represented in the two kingdoms—and bear witness to the fundamental unity of the
nutritive operations in plants and animals.

I shall only be able to indicate in outline the facts and
arguments on which Bernard sought to establish these propositions.

Exterior Digestion.

We all know that the alimentary canal is simply a prolonga-
tion of the external surface ; that the skin is continued, at
either extremity, without a break, into the alimentary mucous
membrane.   Accordingly the processes which take place in the
digestive tube are, strictly speaking, as much outside the body
as if they took place on the surface of the skin.   Upon this in-
ner surface, if I may so call it, are poured out the digestive
juices, charged with the ferments which are the special agents
of the digestive processes.   This is the common condition of
exterior digestion as it occurs in animals—but it is not the
only condition.   Among some of the lowest members of the
animal series a permanent alimentary canal does not exist.   In
the amœba any portion of the exterior is adapted for the recep-
tion of food.   The morsel sinks into a depression formed on
the surface at the point of impact—it is digested in this improvised stomach, and the indigestible portions are expelled
through an improvised anus.

Among plants exterior digestion is a much less prominent
feature than among animals, but examples of its occurrence
and evidence of its importance are not difficult to point out.
In the lowest orders of plants—fungi and saprophytes, which
are devoid of chlorophyll—exterior digestion is probably a
function of prime necessity.   In all likelihood their carbon-
containing food is only absorbed after undergoing a process of
true digestion.   The transformation of cane-sugar by the yeast
plant is a striking example—though a distorted one—of exterior digestion.   Cane-sugar is a crude form of food both to
plants and animals, and requires to be transformed into invert-sugar (a mixture of equal parts of dextrose or grape-sugar
and lævulose or fruit-sugar) before it can be made available for
nutrition.   The yeast plant is no exception to this rule ; and

when placed in a solution of cane-sugar it is under the necessity of transforming that compound into invert-sugar before it can use it for its profit in fermentation. This transformation is effected by a soluble ferment attached to the yeast cell, which can be dissolved from it by water. We shall see later on that a similar ferment exists for a similar purpose in the small intestine of animals—having the same property of changing cane-sugar into invert-sugar.

Even among the higher plants exterior digestion is not quite unknown. The function may be said to be foreshadowed in the excretion of an acid fluid by the rootlets of some plants which serves to dissolve and render absorbable the mineral matters in their vicinity. But genuine and most remarkable examples of this type of digestion occur among the so-called insectivorous plants, of which Mr. Darwin has given so interesting an account. In the sundews, the plant, by a peculiar mechanism provided on its foliage, seizes the insects which fortuitously alight on its leaves. A stomach is extemporised around the prey, into which is poured out a digestive fluid. The prey is digested, and the products absorbed, in essentially the same manner as in the gastric digestion of animals.

Interstitial Digestion.

Both animals and plants lay up reserves, or stores, of food in various parts of their tissues for contingent use, so that if you suddenly withdraw from them their food supplies neither animal nor plant immediately dies—it lives for a certain time on its reserves. But before these reserves can be made available for the operations of nutrition they must first be converted from their inert and mostly insoluble state into a state of solution and adaptability to circulate in the nutritive fluid which constitutes the alimentary atmosphere of the protoplasmic elements. This conversion of inert store-food into available nutriment is brought about certainly in some, presumably in all cases, by the same agents and processes as the digestion which takes place in the alimentary canal of animals ; and it is this identity in the agents and the processes which Bernard insisted on as the proof of the fundamental identity of the two kinds of digestion.

The storing up of food is carried on to a larger extent in the vegetable than in the animal kingdom, owing to the intermittent life of most plants. In their seed, tubers, bulbs, and other receptacles are laid up stores of albumen, starch, cane-sugar and oil—designed primarily for the growth and nutrition of the plant or its offspring—but which are largely seized on by animals and utilized for their food. Owing to their more continuous life animals store up food less than plants. Nevertheless, they accumulate stores of fat in various parts of their body—of animal starch (glycogen) in their livers and elsewhere—and of albumen in their blood. Birds also store :p large quantities of albumen and fat in their eggs.

The transformation of store food has been followed out most completely in regard to starch and its congener glycogen, and cane-sugar. Bernard worked out this subject with marvellous minuteness and success. It has long been known that the transformation of starch into sugar in germinating seeds was effected by diastase ; and that a similar ferment, existing in saliva and pancreatic juice, performed the same office on the starchy food of animals. It has also been proved that the stores of starch laid up in the tuber of the potato and in various parts of other plants are changed at the periods of budding and growth in the same way and by the same agent. Bernard showed that animal starch or glycogen is stored up largely not only in the liver but in a variety of other situations, and especially that it is widely distributed and invariably present in large quantities in embryonic conditions. In juxtaposition with the glycogen is found a diastasic ferment which transforms it into grape-sugar, as it is required for the active operations of growth and nutrition.

The stores of cane-sugar which exist in the beet-root and in the sugar-cane are transformed or digested in like manner into invert-sugar when the plants enter on the second phase of their life—the phase of inflorescence and fructification. Here, again, it has been proved that the converting agent is a soluble ferment—the same ferment which, as alredy mentioned, is attached to the yeast cell—and the same ferment which exists in the small intestine of animals for a similar purpose.

The transformation of store proteids and fats has not been followed out with the same success as that of starch and cane-sugar. But the evidence, as far as it goes, and analogy, point to the conclusion that the stores of albuminous and oily matters contained in the seeds, bulbs, and other receptacles of vegetables are subjected to a digestive process before they are made available for the nutritive operations of the plants, and that the changes thus effected are of the same nature and accomplished by the same agents as those which take place in the digestion of proteids and fats in the alimentary canals of animals.

I have, I think, said enough to show the scope of the evidence and the analogies from which Bernard deduced certain far-reaching generalizations, which I have ventured to summarize in my own language in the following propositions :—

(1) Digestion, or the process by which crude food is changed into available nutriment, is a function or faculty of capital importance in every form of active life.   (2) This function is exercised partly on food brought into proximity with the surface of the organism (exterior, chiefly intestinal digestion) and partly on reserves of food laid up in the interior of the organism (interstitial digestion).   (3) The agents concerned in this function and their mode of action are essentially the same whether the organism be a plant or an animal—and whether the action take place in the interior of the tissues,—or on the general or intestinal surfaces.

General Characters and Properties of the Digestive Ferments.

The essential work of digestion is carried out by a remarkable group of agents, called soluble or unorganised ferments. These are found dissolved in the several digestive juices or secretions, which are thrown out on the path of the food as it travels along the alimentary canal.   The physical and mechanical processes to which food is subjected in the mouth and stomach, are all purely introductory, or preparatory, and are solely intended to facilitate the essential work of digestion, which consists in the action of the digestive ferments on the alimentary principles.

The number of distinct ferments employed in the digestion of the miscellaneous food used by man is not accurately known, but there are at least seven or eight of them.   The accompanying table presents in one view a scheme of the several digestive secretions or juices—and the ferments which they contain—together with an indication of the action of each ferment on the several alimentary principles.

TABLE OF THE DIGESTIVE JUICES AND THEIR FERMENTS.

DIGESTIVE JUICES.	FERMENTS CONTAINED IN THEM.	ACTION ON FOOD MATERIALS.
Saliva....................	Salivary Diastase or Ptyalin..................	Changes Starch into sugar and dextrine.
Gastric Juice	a. Pepsin	Changes Proteids into peptones in an acid medium.
	b. Curdling Ferment ...	Curdles the casein of milk.
Pancreatic Juice ...	a. Trypsin..................	Changes Proteids into peptones in alkaline and neutral media.
	b. Curdling Ferment ...	Curdles the casein of milk.
	c. Pancreatic Diastase..	Changes Starch into sugar and dextrine.
	d. Emulsive Ferment...	Emulsifies and partially saponifies fats.
Bile....................	?	Assists in emulsifying fats.
Intestinal Juice	a. Invertin	Changes cane-sugar into invert-sugar.
	b. ? Curdling Ferment.	Curdles the casein of milk.

An examination of the table shows that a long and complicated series of ferment-actions is required to accomplish the digestion of our food. Starch is attacked at two points—in the mouth and in the duodenum—by two ferments, salivary and pancreatic diastase, which are substantially identical. Albuminous matters are also attacked at two points—in the stomach and in the small intestine—but here the two ferments, pepsin and trypsin, are certainly not identical. The ferment, of which the only known characteristic is to curdle milk, is found in the stomach and in the pancreas—and I think also in the small intestine. The bile is not known to possess any true ferment-action, but it assists, by its alkalescent reaction and by its physical properties, in emulsifying and promoting the absorption of fatty matters. The ferment which transforms cane-sugar, strange to say, is not encountered until the food reaches the small intestine.

The known digestive ferments all belong to the class of soluble or unorganised ferments. They are sharply distinguished from the insoluble or organised ferments, of which the type is yeast, in not having the power of self-nutrition and self-multiplication. All living organisms possess this power either in a dormant (potential) or in an active (kinetic) state. Soluble ferments cannot therefore be said to be alive—but they are exclusively associated with living organisms and take an essential part in their vital operations.

The digestive ferments are all the direct products of living cells and may be regarded as detached repositories of cell-force. They are quite unknown in the domain of ordinary chemistry. Their mode of action bears no resemblance to that of ordinary chemical affinity, and has a distinctively physiological character. They do not derive their marvellous endowments from their material substance. They give nothing material to, and take nothing material from, the substance acted on. The albuminoid matter which constitutes their mass is evidently no more than the material substratum of a special kind of energy—just as the steel of a magnet is the material substratum of the magnetic energy—but is not itself that energy. This albuminoid matter of the ferment may be said to become charged, at the moment of elaboration by the gland-cells, with potential energy of a special kind—in the same way that a piece of steel becomes charged with magnetism by contact with a pre-existing magnet. The potential energy of the ferment is changed into the active form (*i.e.* becomes kinetic) when it is brought into contact with the alimentary substance on which it is designed to act.

The chemical and physical characters of the digestive ferments appear to be tolerably uniform. In composition they resemble proteid substances, and contain carbon, oxygen, hydrogen, and nitrogen in the same or somewhat similar centesimal proportions as albumen. But as not one of them has yet been obtained in a state of absolute isolation and purity, this is, strictly speaking, a matter of inference rather than an ascertained fact.

They are all soluble in water ; they are all diffusible, though with difficulty, through animal membranes and parchment paper. They are also capable (all those I have tried) of passing through porous earthenware by filtration under pressure ; but some of them pass through readily, and others with the utmost difficulty and only in the smallest proportions.

They are precipitated from their watery solutions by absolute alcohol—but unlike other proteids (peptones excepted) they are not truly coagulated by alcohol. When the alcohol is removed the ferments are still found to be soluble in water and to retain their activity unimpaired.

They are coagulated and rendered permanently inert by the heat of boiling water; and when in solution they are coagulated and destroyed by a heat of about 160° Fahr. (71° C.).

[There is a curious point in regard to the proteolytic ferments which requires elucidation. Liquid preparations of these ferments invariably contain a considerable amount of unchanged —that is to say of undigested—albumen, which is precipitated by the addition of nitric acid or by boiling. This is the case both with pancreatic extracts and with solutions of pepsin acidulated with hydrochloric acid. It is evident that this substance cannot be any one of the ordinary forms of albumen— otherwise it would long since have undergone digestion—it would in fact have been transformed into peptone by the trypsin or pepsin associated with it in the solution—and in that condition would of course have been incapable of being precipitated by nitric acid or boiling. All this leads up to the inference that the albuminoid matter which constitutes the organic substratum of pepsin and trypsin is an altogether special form of albumen—and that one of its peculiarities is that it is unsusceptible of the proteolytic transformation which we call digestion. Its relation to ordinary albumens would resemble that of an unfermentable sugar in regard to ordinary sugars. It further suggests itself to one's mind that the undigested remnant which is invariably found as a residuum in the artificial digestions of proteids—and which goes by the name of "dyspeptone"—is not, as has been thought, a bye-product of the digestive process, but that it is simply an admixture of this variety of non-digestible albumen.]

Each digestive ferment has its special correlative alimentary principle, or group of principles, on which alone it is capable of acting. Diastase acts exclusively on amylaceous substances. Pepsin and trypsin act only on the azotised principles—the emulsive ferment of the pancreas is only capable of acting on fatty bodies—the inversive ferment of the small intestine has no activity except on cane-sugar.

The changes impressed on alimentary principles by the digestive ferments are not, chemically speaking, of a profound character—and they affect much more the physical state of these principles than their chemical composition. In the main they are processes of deduplication and hydration—and the result is to render the substances operated on more soluble and more diffusible—to diminish their colloidal state and to make them approach or even to reach the crystalloid state. This does not appear to be an invariable event however. Cane-sugar is a marked exception—it is converted in the small intestine into invert-sugar (a mixture of equal parts of grape-sugar or dextrose and fruit-sugar or lævulose) but invert sugar, though more highly hydrated than cane-sugar, is neither more diffusible nor more soluble.

It does not appear to be absolutely true that all food requires digestion before it can be absorbed. Fat is largely taken up by the lacteals in its unaltered state—except in so far that it is finely divided or emulsified. Grape-sugar (dextrose) is not known to suffer any digestive operation, but to be absorbed unchanged. Perhaps it would be more correct to say that grape-sugar is an article of food predigested for us by the agency of plants.

Although the mode of action of digestive ferments is special and peculiar, the results of that action are not peculiar, but can be obtained in other ways by ordinary chemical forces. By long continued boiling in water—and more rapidly by boiling with acidulated water—starch is converted into dextrine and sugar, and albumen is changed into a substance resembling peptone. The peculiarity of the action of ferments consists in this : that the ferments are able, swiftly and without violence, to produce changes which, by ordinary chemical agencies, can only be produced either by strong reagents or by long-continued and very slow action of weaker reagents. It is interesting to remark that the changes produced in food by digestion are, in their ultimate results, very similar to, if not identical with, those produced by protracted cooking.

[It is becoming evident that the active work which is prosecuted in many quarters on the action of ferments demands the introduction of some new words. The word ferment is still commonly applied to both organised and soluble ferments, although the necessity of referring these groups of agencies to separate categories is universally recognised. Kühne has proposed to designate the soluble ferments as "enzyms," and we may conveniently adopt the word into English, with a slight change of orthography, as "enzymes." May we not also designate the organised ferments as "zymes"? If this suggestion were adopted, a good deal of paraphrase might be avoided by coining from these two roots the cognate words which we are in want of for clear expression and concise description.

The words ferment and fermentation have been so associated from old time with yeast and alcoholic fermentation, that the application of the same words to the processes by which milk-sugar is changed into lactic acid, and alcohol into acetic acid, and to these more complex transformations which we call decomposition and putrefaction, appears strange to us. And yet it is now well known that all these transformations are produced by the action of minute organisms and that they belong strictly to some category as alcoholic fermentation. Still more strange to us is it to apply the same terms to the silent transmutations which take place in the action of diastase on starch and the action of pepsin and trypsin on albumen.

If all organised ferments became known as "zymes," and all soluble ferments as "enzymes," then the process in which zymes are engaged might be called "zymosis"—and the process in which enzymes are engaged might be termed "enzymosis." The action of the former might be described as "zymic," and that of the latter "enzymic." It would also follow that in scientific description the verb to "ferment" would be displaced by the verb to "zymose," or to "enzymose," as the case might be.]

The Preparation of Artificial Digestive Juices.

The study of the digestive ferments has been immensely facilitated by a method first introduced by Eberle. Eberle discovered that an aqueous infusion or extract of the digestive glands possessed the same properties as the natural secretions or juices of those glands. The reason of this is that the glands which secrete the digestive juices contain within them a reserve stock of their respective ferments. Accordingly when the glands are infused in water their reserve stock of ferments passes into solution. These infusions or extracts then constitute artificial digestive juices which operate in a flask or beaker in the same way as the corresponding glandular secretions act in the alimentary canal. Solutions of organic matters are how-ever extremely perishable—they pass quickly into putrefaction. In order to obviate this inconvenience, and to obtain an extract which is always handy for use, various preservative means have been employed. Bernard used carbolic acid—others have used glycerine or common salt. These preservatives, although perfect for the purpose intended, have a pronounced taste which it is impossible to get rid of. I have made a good number of experiments on this point. As the ultimate object I had in view was to obtain a solution which could be administered as a medicine by the mouth—or which could be employed in the preparation of artificially digested food—I sought for a preservative which either had little taste and smell, or one which was volatile and could be got rid of by vaporization. After a good many trials I adopted the three following solutions as on the whole the best suited for the purpose.

I. *Boracic Solution.*

This solution contains three or four per cent of a mixture of two parts of boracic acid and one part of borax. An extract of the stomach or of the pancreas made with this solution keeps perfectly, and has little taste and no smell. For experimental purposes this extract is, I believe, all that can be desired—it is neutral in reaction and is chemically inert. It answers well also for administration by the mouth when the dose does not exceed one or two tea-spoonfuls. But when larger quantities are required for the preparation of artificially digested food, and when food thus prepared has to be used day after day in quantity sufficient to sustain nutrition, larger quantities of boracic acid and borax are taken into the stomach than that organ can always comfortably tolerate.

II. *Dilute Spirit.*

The second solution is water mixed with twelve or fifteen per cent of rectified spirit.* This solution makes a most excellent extracting medium, and the quantity of spirit in it is so small as rarely to be an objection to its use. In the preparation of artificially digested food a final boiling is usually requisite, and in this final boiling the alcohol is dissipated. On the whole, this is the most generally useful of the three solutions.

* This proportion of spirit is not enough for perfect preservation in summer weather. The water should contain 20 to 25 per cent of rectified spirit. The solvent properties of the medium are not in the least deteriorated by this additional proportion of spirit—at any rate in regard to malt-diastase and the pancreatic enzymes.

III. *Chloroform Water.*

Chloroform dissolves in water in the proportion of about one in two hundred—these are the proportions employed in the preparation of the *Aqua Chloroformi* of the British Pharmacopœia. This forms a perfect solvent for the digestive ferments, and its keeping qualities are unrivalled. But though the quantity of chloroform dissolved is so minute (about two and a half drops in a fluid ounce) it communicates a somewhat powerful smell and taste to the solution. This taste and smell are agreeable to most persons, but not to all. It is, however, quite easy to get rid of the chloroform. If the dose to be used be first poured into a wine glass or saucer and left exposed to the air for three or four hours, the chloroform passes off almost entirely in vapour, and leaves behind a simple aqueous solution of the ferments. Or, if the solution is employed for the preparation of artificially digested food, the chloroform, being very volatile, disappears in the final boiling. It is perhaps well to mention that chloroform-water has the property of reducing Fehling's solution, and that this property has to be taken into account in making experiments on digestion which involve testing for sugar.

Alimentary substances fall naturally into three well-marked groups, namely, *Carbohydrates, Proteids,* and *Fats.* I propose to consider the digestion of the three groups in the order here indicated. I have, however, no intention of dealing systematically with these subjects, but rather to take up certain points and questions in regard to which I have myself made observations, or which have a bearing on the preparation of artificially digested food and its administration to patients. I shall treat the digestive transformation of starch in some detail, because this has been worked out almost to completeness, and because it probably furnishes a type which will hereafter be of service as a guide to the study of the more complex problem of the digestion of proteids.

DIASTASIC FERMENTS—DIGESTION OF STARCH.

The importance of starch as an article of human food has, perhaps, scarcely been duly recognised. If we regard the enormous proportion in which the seeds of cereals and leguminous plants and the tuber of the potato enter into our dietary, and the immense percentage of starch in these articles, it is probably not too much to say that fully two-thirds of the food of mankind consists of starch.

In the raw state starch is to man an almost indigestible substance ; but when previously subjected to the operation of cooking it is digested with great facility.

Diastase has only a feeble action on the unbroken starch granule, even at the temperature of the body. In the lower animals, and in germinating seeds, the starch granule is probably attacked in the first instance by some other solvent, which penetrates its outer membranes, and thus enables the diastase to reach and act on the starchy matter contained within. By the aid of heat and moisture in the process of cooking, the starch granule is much more effectively broken up. Its contents swell out enormously by imbibition of water, and the whole is converted, more or less completely, into a paste or jelly or mucilaginous gruel. It is in this gelatinous form exclusively, or almost exclusively, that starch is presented for digestion to man.

The digestion of starch is accomplished partly by the saliva and partly by pancreatic juice, both of which are rich in diastase. Diastase also exists abundantly in the liver, and in smaller quantities in the intestinal juice, in the blood, the urine, and apparently in all the interstitial juices. Diastase from all these diverse sources appears to act substantially in the same manner on starch, changing it by a progressive hydrolysis into sugar and dextrine.

If the action of a fluid containing diastase—say saliva or extract of pancreas—on starch paste be watched, the first effect observed is the liquefaction of the paste and the production of a diffluent solution. This change is effected with great celerity —in two or three minutes the stiff paste becomes a watery liquid. This is evidently a distinct act—and antecedent to the saccharifying process which follows. By operating with small proportions of diastase and large proportions of pure starch paste it is possible to hit on a moment when liquefaction is complete and saccharification is not yet begun. At this moment the solution yields a pure starch reaction, and no reaction of dextrine nor of sugar. The process of saccharification follows immediately on the heels of liquefaction ; and in ordinary manipulations the one process runs into the other.

The speed of the action depends primarily on the proportion of the diastase. By adjusting the proportions of diastase and starch in such degrees that saccharification will be completed in about a couple of hours, the successive steps of the process can be leisurely followed by applying from time to time the appropriate tests.

If you test as soon as liquefaction is complete you get a pure blue with iodine and a slight reaction of sugar with Fehling's solution. In a few minutes the sugar reaction becomes more decided ; and, although you still get a pure blue with iodine in the ordinary way of testing, you will get, by greatly diluting the blue solution and then adding more iodine, a deep violet tint —showing the presence of erythro-dextrine mixed with starch. The next step is the total disappearance of the blue reaction with iodine, and the substitution for it of an intense reddish-brown colouration of erythro-dextrine. Bye and bye the reddish-brown colour is replaced by a yellowish-brown—indicating the preponderating presence of a different kind of erythro-dextrine. Meanwhile the sugar reaction goes on increasing. The next step is the entire disappearance of any kind of coloration with iodine. But the action is still very far from complete—the proportion of sugar goes on increasing for a considerable time after iodine has ceased to tint the solution. At length, however, matters come to a standstill, and the proportion of sugar ceases to increase.

The explanation of this series of reactions is impossible on the old view of the constitution of starch. Until recently it was supposed that the starch molecule was represented by the comparatively simple formula $C_{12}H_{20}O_{10}$, and that under the influence of diastase this molecule was resolved by hydration into two molecules—one of dextrine and one of grape-sugar.

The researches of Musculus and O'Sullivan[*] have shown that this is not a correct account of the transformation. In the first place it was found that the sugar produced was not grape-sugar (dextrose), but another kind of sugar called *maltose*. It was also found that the dextrines first produced, and which were coloured red or brown by iodine, were progressively changed, with simultaneous production of sugar, into a series of dextrines of a lower type, which did not yield any coloration with iodine. To these latter kinds of dextrine the term achroo-dextrines has been applied.

_____
[*] O'Sullivan's papers are published in the Journal of the Chemical Society, from 1872 to 1876. A full account of these researches is given in a paper in the same Journal for September, 1879, by T. H. Brown and J. Heron.

As maltose is now ascertained to be the kind of sugar which is mainly produced in the digestion of starch by diastase, this body assumes a new and considerable importance in physiological chemistry, and it will not be out of place here to give some description of its properties. Maltose is a fermentescible, crystalline sugar of the saccharose (cane-sugar) class, having very little sweetening power, and possessing one atom less water then grape-sugar. Its formula is $C_{12}H_{22}O_{11}$. It possesses more rotatory power on polarized light than grape-sugar, but has considerably less power of reducing cupric oxide. The rotatory power of maltose is $+150$, that of grape-sugar $+58$. The reducing power of maltose is 61 compared to that of grape-sugar as 100. Maltose can be hydrolysed into grape-sugar by prolonged boiling with dilute acids. Malt-diastase does not possess this power, but we shall presently see that the diastasic ferments of the small intestine are able slowly to effect the same change.

The researches of Musculus and O'Sullivan have rendered it necessary to assume that the molecule of soluble or liquefied starch is a composite molecule, containing several members of the group $C_{12}H_{20}O_{10}$—which is to be regarded as the constituent radical of the composite starch molecule. The starch molecule must in the future be represented by the formula $n(C_{12}H_{20}O_{10})$—the value of n not being yet definitely agreed upon.

Two able chemists of Burton-on-Trent, H. T. Brown and J. Heron, have extended these researches, and fully confirmed the main conclusions of Musculus and O'Sullivan. In a recent publication (Journ. Chem. Soc., Sep., 1879) they have for the first time presented a fairly complete scheme of the succession of changes undergone by starch under the action of diastase. These chemists assume that the molecule of soluble starch consists of ten members of the group $C_{12}H_{20}O_{10}$ and that its formula should be written $10(C_{12}H_{20}O_{10})$. This view greatly facilitates the comprehension of the progressive hydrolysis of starch by diastase.

We have seen that starch in the condition of paste or jelly, is distinguished sharply by its physical properties from liquefied or soluble starch. There is therefore in all probability some difference of molecular aggregation between starch in these two states—and it will not be a very bold assumption to suppose that starch in the gelatinous state consists of still more complex molecules than soluble starch—and that several molecules of soluble starch are grouped together to form the molecule of starch in the gelatinous state.

On the ground of these assumptions we may represent the successive steps of the digestion of gelatinous starch by the following series of equations.

The molecule of gelatinous starch is first resolved into its component molecules of soluble starch. The molecule of soluble starch is then resolved by progressive deduplication and hydration into dextrine and maltose by the following succession of steps:—

One molecule of soluble starch $= 10(C_{12}H_{20}O_{10}) + 8(H_2O) =$

1	Erythro-dextrine	α	$9(C_{12}H_{20}O_{10}) +$	$(C_{12}H_{22}O_{11})$	maltose.
2	Erythro-dextrine	β	$8(C_{12}H_{20}O_{10}) +$	$2(C_{12}H_{22}O_{11})$	,,
3	Achroo-dextrine	α	$7(C_{12}H_{20}O_{10}) +$	$3(C_{12}H_{22}O_{11})$	,,
4	Achroo-dextrine	β	$6(C_{12}H_{20}O_{10}) +$	$4(C_{12}H_{22}O_{11})$	,,
5	Achroo-dextrine	γ	$5(C_{12}H_{20}O_{10}) +$	$5(C_{12}H_{22}O_{11})$	,,
6	Achroo-dextrine	δ	$4(C_{12}H_{20}O_{10}) +$	$6(C_{12}H_{22}O_{11})$	,,
7	Achroo-dextrine	ϵ	$3(C_{12}H_{20}O_{10}) +$	$7(C_{12}H_{22}O_{11})$	,,
8	Achroo-dextrine	θ	$2(C_{12}H_{20}O_{10}) +$	$8(C_{12}H_{22}O_{11})$	,,

The final result of the transformation is represented by the equation $10(C_{12}H_{20}O_{10}) + 8H_2O = 8(C_{12}H_{22}O_{11}) + 2(C_{12}H_{20}O_{10})$.
       Soluble Starch.   Water.   Maltose.   Achroo-dextrine.

In order to render this array of equations more easy of comprehension to those who are unaccustomed to read complex chemical formulæ, the transformation may be represented by the subjoined diagram.

We will assume that the molecule of gelatinous starch consists of an aggregation of five molecules of soluble starch, and that the molecule of soluble starch consists of an aggregation of ten groups of the radical $C_{12}H_{20}O_{10}$. Each of these radicals is represented in the diagram by the shaded dots. The open circles represent the atoms of maltose which are set free at each stage of the transformation. The first act is the breaking up of the large molecule of gelatinous starch into its component molecules of soluble starch. Then follows the progressive disintegration of the latter molecules into dextrine and maltose.

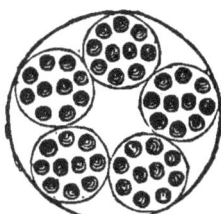

Molecule of
Gelatinous Starch.

Molecule of
Soluble Starch.   + 8 atoms of water.

1 Erythro-dextrine α   maltose.

2 Erythro-dextrine β   ,,

3 Achroo-dextrine α   ,,

4 Achroo-dextrine β   ,,

5 Achroo-dextrine γ   ,,

6 Achroo-dextrine δ   ,,

7 Achroo-dextrine ε   ,,

8 Achroo-dextrine θ   ,,

We must conceive that the energy of the ferment is exercised in gradually pulling asunder the component groups or radicals of the unstable molecule of soluble starch—detaching one after another from the parent molecule—each radical as soon as detached assuming an atom of water and becoming an atom of maltose. At each detachment the parent molecule draws its remaining groups together to form a new kind of dextrine. As the process goes on the dextrine molecule becomes smaller and smaller—that is, contains fewer and fewer component radicals

B

—the higher dextrines giving a red or brown coloration with iodine, but the lower dextrines giving no reaction with iodine.

It is to be noted that after the transformation has reached its final term there still remains a portion of achroo-dextrine unconverted into maltose. Upon this remnant diastase has only a very slow action. The percentage result, when the reaction is completed, gives, in round numbers, 80 parts of maltose and 20 parts of achroo-dextrine. The eight varieties of dextrine indicated in the above table of equations have not all been obtained in the separate state, but there is strong evidence of the existence of at least several of them as distinct bodies.

The account just given of the transformation of starch has been deduced from a study of the action of diastase derived from malt. The question arises—physiologically an important question—whether the action of salivary and pancreatic diastase is identical with that of malt-diastase. The researches of Musculus and V. Mering* give an affirmative answer to this question. These observers found that saliva and pancreatic extract act on starch paste in the same way as malt-diastase, the final products in all cases being achroo-dextrine and maltose, and not dextrose (grape-sugar). At my suggestion Mr. H. T. Brown was good enough to submit the question to a fresh examination in regard to pancreatic extract. His results fully confirm the conclusions of Musculus and V. Mering. He found, however, that there was a slight difference in the results when the action of pancreatic extract and malt-diastase on starch were continued a long time. The pancreatic ferment, in addition to the power, which it shares with malt-diastase, of slowly converting the lowest achroo-dextrine into maltose, exhibited a power of slowly changing maltose into dextrose (grape-sugar) which is not possessed in any degree by malt-diastase. Mr. Brown also informs me that there is in the small intestine a ferment which possesses similar properties. †

The respective shares of Saliva and Pancreatic Juice in the Digestion of Starch.

The respective shares of saliva and pancreatic juice in the digestion of our farinaceous food is probably variable and perhaps not quite identical.

*Maly's Jahres-bericht für Thier-Chemie for 1878, p. 40.

† See a paper by Brown and Heron on the Hydrolytic Ferments of the Pancreas and Small Intestine in the Proceedings of the Royal Society for 1880, p. 303.

As all our farinaceous food is eaten after being cooked, the starch in it is more or less completely gelatinized; it is, therefore, probable that one of the chief uses of salivary diastase in man is to liquefy starch jelly. A very brief contact suffices for this, and it is manifest that the accomplishment of this change is an important advantage in the subsequent operations in the stomach. Our gruels, blanc-manges, puddings, and similar farinaceous dishes owe their thick pasty condition to starch in the gelatinous state, and nothing can be imagined more resistent to the rapid permeation of a meal by the gastric juice, and to the pulping of it into a uniform chyme, than the presence of coherent masses of starch paste. If the saliva performed no other service than this it would furnish an important aid to the digestion of a meal.

There has been considerable dispute as to whether, and how far, the saccharification of starch goes on in the stomach. My own observations lead to the conclusion that this depends on the degree of acidity of the contents of the stomach; and it is known that this varies within very wide limits. When a meal is swallowed it takes some time for the gastric juice to permeate the mass, and the acidity of the gastric contents is for some time feeble. As digestion proceeds the contents of the stomach tend to become more and more acid. This is a point which each one can observe for himself. The stomach is by no means reticent of its doings. The posseting which we see in infants goes on in a less degree in the adult; and we are perforce made aware, sometimes inconveniently so, by our palates, of the ascending scale of acidity in the stomach as digestion advances. Saliva acts energetically in neutral and in slightly acid media, but its activity is checked and finally arrested when the acidity becomes pronounced. When digestion is proceeding comfortably and normally a certain interval elapses before the acidity of the stomach becomes considerable, and during this interval the salivary diastase continues active, and has time to accomplish a good deal of work. But we must remember that our farinaceous food is, for the most part, not in the most favourable condition for rapid digestion. It is not generally in a state of mucilage— but in the form of a solid paste as in bread, puddings, and pastry. A good deal of it too is imperfectly cooked. Consequently, the larger part of our starchy food reaches the duodenum still unchanged, or only partially changed, and this larger part of the work is consummated by the pancreatic juice in the alkaline medium of the small intestine. I shall have to return to this point in speaking of gastric digestion.

It has been noted as curious that the saliva of man possesses more diastasic power than that of almost any other animal. Among the herbivora, which are such large consumers of starch, the saliva has comparatively little diastasic power; and in some, as in the horse, it is almost or altogether wanting. I apprehend that this is due to the fact that man alone has learnt to cook his starchy food, and that the diastasic power of his saliva has become developed with the opportunity for its exercise. Diastasic power would be thrown away in the saliva of the horse, because he eats his food in the raw or uncooked state, and saliva is almost without action on raw starch.

When can Starch be said to be fully Digested?

Seeing that in the digestion of starch a number of intermediate products are evolved—the question arises when can the digestion of starch be said to be accomplished? Is maltose the only product absorbed, or are not the dextrines, especially the achroo-dextrines, also absorbed? The dextrines, even those coloured by iodine, are highly diffusible, and pass freely through parchment paper in dialysis. In this respect they contrast strongly with starch jelly, and even with liquefied (or soluble) starch, both of which are undialysable. It seems not improbable that the lower dextrines are largely absorbed. Because if we follow the history of starch after it has been transformed by digestion, and absorbed, we are confronted with the remarkable fact that after absorption the products of starch digestion, or at least a large portion of them, undergo a reconversion in the liver into a substance closely resembling undigested starch. Glycogen, in its essential features, is an exact counterpart of soluble starch. It forms an opalescent solution in water; it is undialysable, and it is transformed by diastase into dextrine and sugar.

It appears reasonable to suppose that it would be an advantage to the economy if that portion of our starchy food which is destined to be stocked in the liver as glycogen, should be absorbed at an early period of the digestion, because the less removed the digested product is from starch at the moment of absorption the fewer steps it will have to retrace in recovering the amylaceous state after absorption.

The annexed diagram will render my meaning clear. It represents, on one side of the dividing line of the absorbent membrane, the descending steps of the digestive process—and on the other side the ascending curve of the reconstructive process.

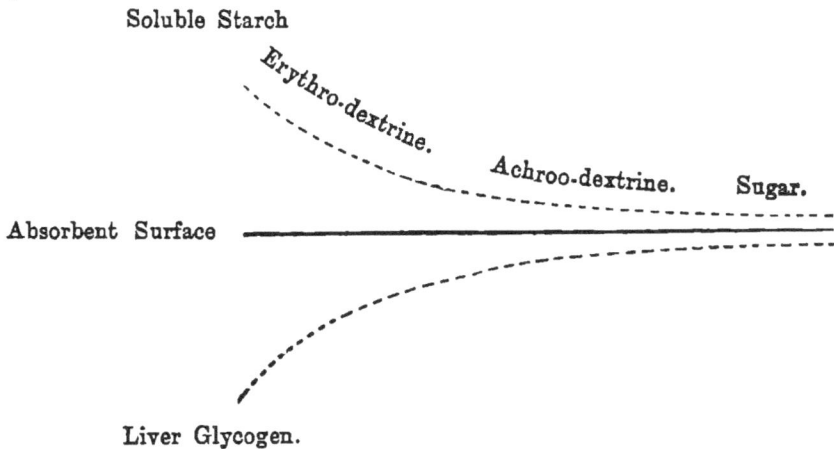

Soluble Starch

Erythro-dextrine.

Achroo-dextrine.　　Sugar.

Absorbent Surface

Liver Glycogen.

It is not necessary to suppose that the ascending steps of the reconversion are identical with the descending steps of digestion, but it is probable that they are fundamentally alike, seeing the close similarity of the products at the two ends of the journey. At any rate there is no warrant in the present state of knowledge for the opinion that sugar is the only absorbable product of starch digestion.

Absolute Energy of Diastase.

The notion that the energy of diastase is not consumed in action seems, on *a priori* grounds, to be quite untenable—such a notion contravenes a general principle in physics that energy in performing work is expended and finally exhausted. It is easy to show experimentally that diastase is no exception to this rule. Payen and Persoz estimated that malt-diastase was able to convert two thousand times its weight of starch into sugar. My own experiments with extract of pancreas indicate a much higher power than this. The following experiment illustrates at the same time the enormous diastasic power of pancreatic diastase and the fact that this power is strictly limited.

A quantity of starch mucilage was prepared, containing one per cent of pure potato starch—100 cubic centimeters of this mucilage contained exactly one gram of dry starch. The pancreatic extract employed was prepared in the following manner :—Fresh pig's pancreas, freed from fat, was rubbed up with an equal weight of fine sand until it became a smooth uniform pulp. This pulp was spread out very thinly on sheets of glass, and allowed to dry in the open air for a fortnight. It was then scraped off with a knife and formed a rough shreddy sort of powder. 125 grams of this mixture of pancreas and sand were infused, at the temperature of the room, in 1,000 cubic centimeters of saturated chloroform water, with a little more chloroform added to ensure against decomposition. The mixture was allowed to stand for four days with occasional agitation, and the product was then filtered clear through paper. The extract of pancreas thus prepared proved a very serviceable preparation, and most of my observations on panereatic digestion were made with it.

This extract was found to be so extremely active that it was necessary to dilute it largely in order to bring the quantities of starch operated on within due compass. Accordingly a dilution was made of one cubic centimeter of the extract in 1,000 cubic centimeters of water. Five numbered phials were then severally charged with 100 cubic centimeters of the prepared starch mucilage—so that each phial contained exactly one gram of dry starch. One cubic centimeter of the diluted pancreatic extract was added to phial No. 1—two cubic centimeters to No. 2— four cubic centimeters to No. 3—six cubic centimeters to No. 4—and eight cubic centimeters to phial No. 5. The phials were then corked and placed in the warm chamber, where the temperature was steadily maintained by a Page's regulator at 100°F. (38°C.)

At the end of twenty hours the contents of the phials were examined. All of them were perfectly transparent, and had entirely lost the opalescent appearance of the original starch mucilage, and not a vestige of sediment existed in any of them. The following reactions indicated the progress of the transformation.

No. 1 gave an intense blue colorotion with iodine—and when the blue solution was largely diluted and more iodine added, it developed a violet tint which showed the presence of erythrodextrine—it also reduced the cupro-potassic solution freely.

No. 2 gave a strong blue reaction with iodine, and by diluting and adding more iodine, the colour changed to a deep claret-red, indicative of abundance of erythro-dextrine. This and all the rest gave a strong sugar reaction with Fehling's solution.

No. 3 yielded no blue reaction with iodine, but an intense port-wine coloration of erythro-dextrine.

No. 4 gave no blue reaction with iodine, and only the faintest possible brown coloration with that reagent, showing only traces of erythro-dextrine.

No. 5 exhibited not a vestige of reaction with iodine. It contained neither starch nor erythro-dextrine, but it yielded a strong sugar reaction.

The transformation of No. 5 might be regarded as complete, but the rest still contained starch or erythro-dextrine, or both. Nos. 1, 2, 3, and 4 were restored to the warm chamber, and re-examined at the expiration of seven hours. No. 4 no longer gave the slightest reaction with iodine, but Nos. 1, 2, and 3 showed only slight signs of further alteration, and were returned to the warm chamber.

At the end of 48 hours from the commencement of the experiment, Nos. 1, 2, and 3 were examined again.

No. 1 showed a strong blue coloration with iodine, and also a strong reaction of erythro-dextrine.

No. 2 no longer gave any blue tint with iodine, but it exhibited an intense erythro-dextrine reaction.

No. 3 only gave a yellowish brown reaction with iodine of moderate intensity.

After a further sojourn in the warm chamber of seventy hours the contents of the three phials were not found to be sensibly altered—they gave exactly the same reactions as before. It was evident that in these phials the diastasic action had run its course to an end within the period of 48 hours, and that the solutions had then come to a state of rest—the ferment had liberated all its energy—the limits of its power had been reached—and the task allotted to it was left unfinished. Nevertheless it had accomplished an amount of work which, considering its infinitesimally minute mass, appears marvellous. We shall now endeavour to measure approximately the amount of this work as indicated by the above experiments.

The original pancreatic extract, when evaporated to dryness in a water-bath, was found to leave a residue of 1·5 per cent of organic matter. This organic matter included, besides diastase, a quantity of proteolytic ferment (trypsin) and a certain quantity of the milk-curdling ferment. It also included a certain quantity of digested proteid matter—for in making an extract of the pancreas there is always accomplished some self-digestion of the glandular tissue.

Taking into account these various admixtures, it would appear a very liberal allowance to estimate the diastasic ferment as amounting to one-fourth of the total organic matter. This would give us for the original extract a proportion of diastase in round numbers of 0·4 per cent, and for the diluted extract of 0·0004 per cent.

The proportion of diastase added to phial No. 4 seems to have hit off with precision the limit of quantity required to transform one gram (15·5 grains) of starch in 48 hours at a temperature of 100° F. (38° C.). The amount of diluted extract added to this phial was 6 cubic centimeters, and on the basis of the above estimate this represents a quantity of net diastase amounting to 0·00024 gram. This yields us by an easy calculation the astounding result that pancreatic diastase is able to transform into sugar and dextrine no less than 40,000 times its own weight of starch.*

. The speed at which a given quantity of starch is transformed by diastase depends essentially on the proportion of ferment brought to act upon it. In the above experiments the proportion of diastase was very minute in comparison with the amount of starch, and the action went on slowly for forty-eight hours. But if we reverse these proportions and mix a small amount of starch with a large amount of diastase the transformation is instantaneously accomplished.

If a test-tube be half filled with an active extract of pancreas and a few drops of starch mucilage be quickly shaken therewith, you cannot detect the reaction of starch or dextrine in the mixture, however prompt you may be with the testing—the transformation has followed on the admixture as instantaneously as the explosion of the charge follows the fall of the trigger. Between these extremes there are all gradations. This mode of action differs entirely from what is seen in the operation of ordinary chemical affinity. If you add a drop of

---

* Marvellous as these numbers are, Mr. Horace Brown (whose joint paper with Mr. Heron, already referred to, has been my chief guide in trying to work out the theory of starch digestion) informs me in a private communication that he has arrived at numbers still more wonderful in estimating the transforming power of malt-extract on the higher dextrines.

acid to an excess of alkali, the acid is instantly neutralized and the action comes to an end ; and, conversely, if you add a drop of alkali to an excess of acid, the action is equally instantaneous ; the affinity of the two bodies for each other is a mutual affinity. But this is not the case with the action of diastase on starch. The starch appears entirely passive in the process ; all the energy is on the side of the diastase, and this energy can only be liberated gradually. There is something in this strikingly suggestive or reminiscent of the action of living organisms. To illustrate my meaning, let us compare the particles of the ferment to a band of living workmen whose function is to scatter little heaps of stones. If the heaps are few and the workmen many, all the heaps will be scattered at once, and the energy of the workmen will still remain sensibly unimpaired. But if the heaps are millions and the workmen hundreds, and if the workmen are doomed to labour on until they fall exhausted at their task, the scattering of the heaps will go on for a comparatively long time and the process of exhaustion will be a gradual one.

I may here mention that the diastasic ferment does not exist in the saliva and pancreatic juice of young sucking animals—except in minute proportions. Its quantity increases when the teeth are cut. In the human infant diastase does not appear to exist in sufficient abundance to digest starchy matters effectively until about the sixth or seventh month. Until this period it is therefore not advisable to administer farinaceous food to infants.

DIGESTION OF CANE-SUGAR—INVERSIVE FERMENT.

Bernard* first called attention to the fact, already mentioned, that cane-sugar (saccharose) required digestion both in animals and plants before it could be used in nutrition. The cane-sugar stored up in beet-root and in the sugar-cane is changed by ferment-action into invert-sugar before it is permitted to circulate in the sap, and take part in the nutritive operations of the plant. He found also that an analogous transformation was requisite before cane-sugar could be assimilated by animals. He states that when cane-sugar is injected into the blood it circulates therein as an inert body, and is in

* Claude Bernard—*Leçons sur les phénomènes de la vie*, T. II., p. 36. Paris, 1879.

no degree used as nutriment by the tissues, but is eventually
entirely removed, unchanged, with the urine. Cane-sugar is,
however, an important article of food, and is consumed by us
in large quantities every day. And we know that when thus
consumed it does not behave like an inert matter—circulating
awhile in the blood, and then being eliminated by the kidneys
as a waste product. It is evidently absorbed and assimilated,
and must therefore, somewhere or other, be transformed or
digested in animals as it is in plants. Reasoning this way,
Bernard sought for an inversive ferment for cane-sugar in the
alimentary tract ; and after searching in the saliva, in the
stomach, and in the pancreas in vain, he at length discovered
it in the small intestine. In the small intestine he found that
cane-sugar was transformed into invert-sugar, and by a similar
ferment with that destined for analagous purposes in yeast, in
beet-root and in the sugar-cane.

The transformation of cane-sugar into invert-sugar is represented by a very simple equation :—

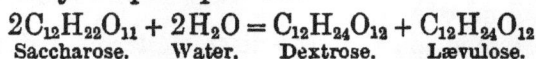

$$2\,C_{12}H_{22}O_{11} + 2\,H_2O = C_{12}H_{24}O_{12} + C_{12}H_{24}O_{12}$$
Saccharose.   Water.   Dextrose.   Lævulose.

The inversive ferment was detected by Bernard in the small
intestine of dogs, rabbits, birds, and frogs. Balbiani found it
in the intestine of the silk-worm. It was recognised by
myself in an extract of the small intestine of the pig, the fowl,
and the hare. It does not exist in the large intestine.

But although my observations on this subject coincided in
the main with those of Bernard, I noted two points which I
think merit further attention. The first was that while a piece
of small intestine infused in water yielded a mixture which was
capable of inverting cane-sugar, the same infusion when filtered
through paper until it was perfectly clear had no such power.
It seemed as if the inversive ferment did not pass freely, if at
all, into true solution, but remained attached to some of the
formed elements contained in the intestine. The second point
I noticed was the extreme slowness of the action. When cane-
sugar was added to the unfiltered infusion of intestine, and the
mixture maintained at blood heat, it generally took a couple of
hours before a reducing effect with the copper test could be
obtained. Both these circumstances reminded one of the action
of formed ferments, and I could not help thinking that there
was here something which required clearing up at some future
time.

LECTURE II.

SUMMARY.—Pepsin and Trypsin and the Digestion of Proteids— Nature of the Acid of Gastric Juice—Effect of Gastric Juice on the Salivary and Pancreatic Ferments—Digestive Proteolysis— Characters of Peptones—Comparison of the action of Pepsin and Trypsin—Peptogens—The Milk-curdling Ferment—The Emulsive Ferment and the Digestion of Fats.

PEPSIN AND TRYPSIN—DIGESTION OF PROTEIDS.

Various kinds of albuminous or proteid substances are used by mankind as food. The most important of these are muscular flesh, the casein of milk, and egg-albumen from the animal kingdom—and gluten, albumen and legumin from the vegetable kingdom.

Proteids are attacked by the digestive ferments at two points in the alimentary canal, by pepsin in the stomach and by trypsin in the small intestine. Between these two acts of digestion there is a complete break in the duodenum, owing to the abrupt change of reaction from acid to alkaline which occurs at that point.

Gastric digestion is, in all creatures, an essentially acid digestion ; but the most varied opinions have been entertained as to the nature of the acid. It has been supposed in turn to be hydrochloric acid, lactic acid, acid phosphate of lime, butyric, and even acetic acid. Much of this confusion has arisen from not distinguishing between the acid of pure gastric juice as secreted into an empty stomach, and the acid of the gastric contents during the digestion of a meal. A good deal of light has been thrown on this subject by the recent researches of C. Richet.* This observer found himself in exceptionally advantageous circumstances for the study of the gastric juice. He had under observation a young man on whom Verneuil had successfully performed the operation of gastrotomy for the relief of impermeable stricture of the œsophagus, the result of swallowing inadvertently a quantity of caustic potash. The complete occlusion of the œsophagus enabled Richet to obtain and examine the gastric juice in a pure state, uncontaminated with saliva. All food had to be administered through the fistulous

* Du Suc Gastrique, by Ch. Richet. Paris, 1878.

opening left after recovery, and the observer could at any moment—as in the famous case of Alexis St Martin—withdraw portions of the gastric contents for examination. Richet also took advantage of the new method of separating the various organic and mineral acids from one another made public by Berthelot in 1872.*

Berthelot had found that if you shook up an aqueous solution of any acid with ether, and then allowed the two fluids to separate, that a part of the acid passed into the ether, and that the remainder clung to the water, and that the ratio between these two parts was a constant quantity. He called this ratio the *coefficient of partage*, and found that its value was a fixed characteristic for each particular acid. Solutions of the mineral acids were found to yield nothing, or almost nothing, to ether when agitated with it—but organic acids were found to pass into the ether in considerable, though very variable, quantities—the proportion varying in a constant ratio according to the nature of the acid.

Testing pure gastric juice, uncontaminated with food or saliva, by this method, Richet found that almost all the acid was retained by the water, and that only a small proportion—about one in twenty-two—passed into the ether. This showed that the acid of pure gastric juice was almost entirely a mineral acid, with only a minute admixture of organic acid. The organic acid (tested by the same method) exhibited a coefficient of partage closely corresponding with that of sarcolactic acid. The nature of the mineral acid was determined by a method similar to that employed by C. Schmidt, and yielded the same result—namely, undoubted evidence that the mineral acid of pure gastric juice was hydrochloric acid. But Richet found that this mineral acid did not behave in the presence of salts of the organic acids quite in the same way as free hydrochloric acid does. The observations of Berthelot had shown that when free hydrochloric acid is added to a solution of acetate of soda, or other similar organic salt, the mineral acid attaches the base entirely to itself, and throws the whole of the organic acid free into solution—so that the mixture when tested by the method of coefficient of partage behaves exactly like an unmixed solution of the organic acid. When pure gastric juice was put to this test it was found that, although it liberated the organic

* *Ann. de chimie et de physique*, 4e Serie, t. xxvi., p. 396.

acid largely, it did not do so to the same extent, by a good deal, as if the mineral acid contained in it had been entirely free. It behaved, rather, as hydrochloric acid does when united with a feeble organic base, such as leucin or peptone—and, on the ground of what appear to be very careful experiments, Richet came to the conclusion that this feeble base was probably leucin, derived from the gastric mucus, and that the acid of pure uncontaminated gastric juice was hydrochloric acid in loose combination with leucin.*

Richet next proceeded to examine the free acid actually existing in the stomach during the digestion of a meal in his patient with gastric fistula. He found, as might have been expected, that this differed from the acid of pure unmixed gastric juice—in this respect, namely, that it contained a very much larger proportion of organic acids, in comparison to the mineral acid, than pure gastric juice. It was evident that the mineral acid had to a large extent seized upon the bases of the acetates, malates, lactates, butyrates, and other organic salts always present in the food, and had set free their organic acids. The real work of digestion then, so far as the acid constituent is concerned therein, is largely performed by various organic acids thus set free from the articles of food which are undergoing digestion.

Richet found that the acidity of the contents of the stomach during digestion, although it varied through considerable limits, had a marked tendency to maintain the normal average. If acid or alkali was added to the digesting mass the mean was presently restored automatically—the stomach in the former case ceasing to secrete acid, and in the latter case secreting an increased quantity of acid.

The Effect of Gastric Juice on the Salivary and Pancreatic Ferments.

The observations of Berthelot on the power of a mineral acid to set free the acids of organic salts, and to take their place with the bases—and the further observations of Richet, showing that

---

* There are other proofs that the acid of the gastric juice is not free hydrochloric acid. It was shown by Bernard that free hydrochloric acid, even in the proportion of one in a thousand of water, dissolves oxalate of lime—but gastric juice possesses no such power. Blondlot likewise proved that gastric juice does not decompose carbonate of lime—whereas the feeblest dilutions of free hydrochloric acid do so.

the acids actually free in the gastric contents during digestion were organic acids, have led to the re-examination of a point of some importance, namely—as to whether the salivary and more especially the pancreatic ferments were or were not destroyed by the acid contents of the stomach.   The matter has a practical interest in this way.   If the gastric acid destroys these ferments it is evidently useless to administer pancreatic preparations by the mouth during digestion, because they would be rendered inert by the acid contents of the stomach.   On the other hand, if they are not destroyed in passing through the stomach, but merely lie dormant and recover their activity in the alkaline medium of the small intestine, then we can administer pancreatic preparations during digestion with every prospect of their passing uninjured through the pylorus and proving useful in assisting digestion in the small intestine.

A series of experiments bearing on this question were submitted by T. Defresne to the Académie de Médecine and the Académie des Sciences toward the close of 1879, and have attracted some attention.   On the ground of these experiments Defresne arrived at the following conclusions—namely, that saliva continued its action on starch in the stomach without interruption—that the pancreatic ferments in like manner preserve their activity in the presence of the gastric acid—that the acids of the chyme, being organic acids, did not really destroy these ferments, but merely reduced them to a state of temporary inertness; so that when the acidity of the chyme was neutralized in the duodenum, they recovered their powers and exhibited undiminished activity both on starch and proteids.

As the question had a direct bearing on the medicinal administration of pancreatic preparations, and indirectly on the administration of malt-diastase and malt-extract, I thought it desirable to repeat some of Defresne's experiments and to put the question raised by these experiments to the test in other ways.

One of Defresne's (apparently) most convincing experiments was the following, which I give nearly in the words of the abstract of the paper published in the Proceedings of the Académie de Médecine for Nov. 4th, 1879.   When 20 grams of dilute hydrochloric acid, having twice the acidity of the normal chyme (which is estimated as 2 per 1000 of HCl), are mixed with 20 grams of egg-albumen—what follows?   The acidity of the medium is no longer due to free hydrochloric acid but to the lactic and phosphoric acids of the white-of-egg which have been set free.   In presence of these acids pancreatine may be digested for two hours in the warm chamber with impunity. And if, at the end of this period, the acidity of the mixture be neutralized, digestion is accelerated and the pancreatine pepto- nises 38 times its weight of albumen.

I repeated this experiment in the following manner : 40 grams of boiled and chopped white-of-egg were mixed with 40 cubic centimeters of dilute hydrochloric acid of the strength of 4 per 1000.   This mixture was subjected to a preliminary digestion of two hours in the warm chamber at 40°C.   The object of this preliminary digestion was to allow the hydro- chloric acid a sufficient time to seize on the bases, and to set free the organic acids of the white-of-egg.   At the end of this time, 5 cubic centimeters of an active extract of pancreas were added to the mixture.   A second experiment was arranged in exactly the same way, except that filtered saliva was substituted for extract of pancreas.   The mixtures were kept in the warm chamber for a further period of one hour, and were then filtered and carefully neutralized.   On testing the neutralized filtrates, I obtained approximatively the same results as Defresne.   The diastasic and proteolytic ferments of the pancreatic extract were found active, but not so active by a good deal as if the extract had been diluted to the same extent with simple water. In the second experiment with the filtered saliva, the ptyalin had preserved its activity quite unimpaired.

These experiments, however, involve a fallacy which vitiates the deductions intended to be drawn from them.   White-of-egg has a highly alkaline reaction, and although the acid used in the experiment possessed double the strength of the normal gastric juice, it was found that the mixtures, at the end of the two hours' preliminary digestion, had only a comparatively feeble acidity—in fact, only one-seventh of the normal acidity of the gastric contents.   It has long been known that the sali- vary and pancreatic ferments are able to resist a feeble acidity, but the question really at issue is : can these ferments resist the average acidity of the contents of the stomach, when, more- over, this acidity is still rendered more destructive to them by the presence of pepsin ?

I tested the question in this way.   A distinction is drawn, and rightly, between acidity due to free hydrochloric acid, and a similar degree of acidity due to an organic acid.   Now, lactic acid is a typical organic acid, and it is also an acid which is often, if not always, present in the contents of the stomach during digestion. I prepared a solution of lactic acid corresponding in saturating power to the normal gastric acid (2 per 1000 HCl.) To 50 cubic centimeters of this dilute lactic acid I added 5 cubic centimeters of a solution of pepsin and 5 cubic centimeters of an active extract of pancreas. I prepared a second similar experiment, but substituted filtered saliva for pancreatic extract. The mixtures were then placed in the warm chamber for one hour. At the end of this period the solutions were exactly neutralized and tested. They were both found to be absolutely inert. Not a vestige of amylolytic nor proteolytic power had escaped destruction.

I had an opportunity of trying the same question in a still more satisfactory way. While I was examining the throat of a patient suffering from an ailment which did not affect his general health, a portion of the contents of the stomach was ejected, and fortunately caught in a clean vessel. This was immediately filtered, and about 10 cubic centimeters of clear acid solution were obtained. The period of digestion was three hours after breakfast. One half of this was devoted to testing its saturating power. It was found to possess an acidity very nearly corresponding with that of normal chyme. To the remaining portion five drops of extract of pancreas and five drops of filtered saliva were added, and the mixture was placed in the warm chamber for one hour. At the end of this time it was exactly neutralized, and divided into two equal portions. One portion was tested with a drop of starch-mucilage, and found to be absolutely devoid of amylolytic power. The other portion was added to an equal volume of milk rendered slightly alkaline with bicarbonate of soda, and was then placed in the warm chamber. Not the slightest digestive action was produced on the milk in twelve hours.

I may mention that in the above experiments I used milk as the test of proteolytic activity. I had become thoroughly familiar with the behaviour of milk with pancreatic preparations during a long course of observations, and was therefore able to detect the slightest signs of pancreatic action.

With this evidence before me, I am unable to accept the conclusions of Defresne and others in Paris who allege that saliva and pancreatic preparations can resist the normal acidity of the stomach in full digestion; and who recommend the administration by the mouth of pancreatic preparations during the period of chymification. I will in my next lecture point out the time, and the method in which these preparations may, I believe, be administered by the mouth with some prospect of success.

Digestive Proteolysis.

The changes undergone by albuminoid substances in digestion are still very imperfectly understood. It is however known that the chief end-product of the transformation is peptone. It is also known that between any native proteid—egg-albumen, muscle-fibrin, casein, or legumin—and the end-product peptone there are intermediate grades and bye products, which have hitherto proved difficult to define and isolate. The constitution of the proteid molecule is still unknown to chemists. That it is a highly complex aggregation is certain; and it can scarcely be doubted that the way to a better knowledge of its constitution lies in a persevering study of the action on it of the digestive ferments. It has been already seen that in the case of starch the action of diastase furnished the key to the constitution of the starch molecule—and, similarly, it is not unreasonable to expect that the mystery of the proteid molecule will be finally solved by a study of the action on it of pepsin and trypsin. Attention has hitherto been too exclusively directed to peptic digestion, which is complicated by the inter-action of an acid. Pancreatic digestion is, in this respect, a simpler process—inasmuch as it requires the interference neither of an acid nor an alkali, but is a reaction, pure and simple, between the ferment and the proteid. This, however, is a question of the future.

Characters of Peptones.—The end-product peptone has been fairly isolated and its characteristics defined. If the filtered product of a pancreatic digestion of egg-albumen be evaporated to dryness in a watch-glass at a temperature of 104°Fahr. (40°C.) there remains a glassy straw-coloured residue resembling dried gum. With the point of a pen-knife this can be chipped off in shining scales, which may be easily reduced to a fine whitish powder—this is nearly pure peptone. This substance is extremely soluble in water: and its solutions even when highly concentrated by evaporation betray no jellying, and no viscosity, but continue diffluent almost to the moment of desiccation—only just before drying up do they assume a slightly syrupy consistence. When the latter point is approached the

C

solution deposits beautiful white crystalline sheaves of tyrosin and spheres of leucin. From the constancy with which these crystalline bodies make their appearance in pancreatic digestion it may be inferred that they constitute an essential portion of the final products of the tranformation.

The reactions of peptone are mostly of a negative character. Its solutions give no precipitation with nitric acid, nor with boiling, nor with ferrocyanide of potassium and acetic acid. The behaviour of peptone with alcohol is peculiar. When a strong peptone solution is dropped into absolute alcohol, the peptone is precipitated as a white sediment, but it is not truly coagulated into an insoluble modification, as all the other proteids are, for when the alcohol is removed the deposit is found to have preserved its solubility in water—even after prolonged contact with the alcohol. Solutions of peptone are precipitated by those metallic salts which throw down other proteids, also by tannin when the solutions are neutral. When the solutions are rendered alkaline, the cupro-potassic test (Fehling's solution) added in very small quantity produces a rose-red coloration, whereas other proteids produce a violet tint. Physiologically, by far the most important reactions of peptone are its extreme solubility in water and its diffusibility through organic membranes. With regard to the latter point contradictory statements have been made by different investigators. Otto Funke rated the diffusibility of peptone through the membrane of the small intestine higher than that of common salt. On the other hand, V. Wittich concluded that peptones did not pass through parchment paper more rapidly than unaltered albumen. My own observations support the view that peptone is incomparably more diffusible through parchment paper than the native proteids. In the case of milk the effect of digestion is very marked in regard to its behaviour in the dialyser. When milk was dialysed for 48 hours into twice its volume of water, not a trace of proteid matter could be detected in the diffusate; but when milk had been previously digested for a couple of hours with pancreatic extract, an abundant reaction with tannin and Fehling's solution was obtained after dialysing it for eight hours.

My results with egg-albumen were equally striking. I prepared a solution by agitating white-of-egg with nine times its bulk of water, and straining the product through muslin. When this solution was dialysed into twice its volume of water for thirty-two hours it yielded absolutely no reaction with tannin nor with the cupro-potassic test. But when the same solution was previously digested, either with pepsin and acid—or with pancreatic extract—and then dialysed, it gave in five hours a slight but distinct reaction, both with tannin and with Fehling's solution, and in sixteen hours a most abundant reaction.

The theoretical views held by physiologists in regard to the intimate nature of the transformation undergone by proteids in the presence of pepsin and trypsin are still unsettled, but recent opinions converge towards the idea that the process is, in the main, one of progressive deduplication with hydration, similar in type to the transformation of starch by diastase. This view receives a positive support from the recent analyses of Henninger. By operating on highly purified albumen, fibrin, and casein, Henninger* succeeded in obtaining peptones in a state of great purity. An analysis of peptones so obtained indicated that they contained less carbon and nitrogen, and proportionally more hydrogen, than their original proteids. The differences were, it is true, small, but they pointed distinctly in the direction of the conclusion that there was a fixation of the elements of water by the proteid in the course of its transformation into peptone.

Henninger also believes that he has succeeded in reversing the process, and in reproducing an albuminous substance from peptone by operating on it with dehydrating agencies. He found that when fibrin-peptone was heated with anhydrous acetic acid at 80° C., or was maintained for an hour at the temperature of 160° to 180° C., it yielded a body which agreed closely in its reactions with syntonin.

The intermediate products which are generated in the course of proteid digestion, and stand between the original proteid and the end-product peptone, are very imperfectly understood. The laborious researches of Meissner and Kühne have shown that several such intermediate bodies are produced, but these have not yet been adequately isolated and defined.

Comparison of the Action of Pepsin and Trypsin.

The action of pepsin and trypsin, although similar in the main results, is certainly not identical. There is a markedly larger production of leucin and tyrosin in tryptic than in peptic

* A. Henninger, "De la Nature et du Rôle Physiologique des Peptonés." Paris, 1878.

digestion. Moreover, the action of the two ferments on different proteids appears to vary both in character and in energy. Milk is much more readily digested by pancreatic extracts than by artificial gastric juice; but in the case of egg-albumen the advantage lies decidedly with the gastric juice. The study of the digestion of egg-albumen by the two methods yielded some interesting results. I employed for this purpose a dilution of egg-albumen (obtained from *fresh-laid* eggs) with water, in the proportion of one in ten. This remains uncoagulated after being boiled in the water-bath, and furnishes a favourable medium for studying the digestion of albumen—more favourable in some respects than the chopped boiled white-of-egg usually employed. It permits the ferment to be at once brought into uniform and intimate contact with the particles of the albumen, thus obviating the irregularity and want of constancy which necessarily attends the operation of a solvent acting on solid pieces of variable size, which can only be attacked progressively from their surfaces. In the raw state this solution is digested with extreme slowness by artificial gastric juice; and pancreatic extract is nearly inert on it; but after being boiled it is attacked with energy by both the gastric and the pancreatic ferments.* When the boiled solution was treated, in the warm chamber, with pepsin and hydrochloric acid, the transformation of the albumen went on rapidly and without interruption to its close. In the earlier periods of the action the mixture gave a dense precipitate with nitric acid and with ferrocyanide of potassium, but this precipitation became progressively less and less pronounced, until at the end of two or three hours these reagents only produced a slight haze. The albumen was now completely digested, or at least as nearly so as could be reached, for this remnant of a reaction persisted even after a further digestion of twenty-four hours.

When the same solution was treated with pancreatic extract the progress of events was different. For an hour or two (the time varying with the quantity of extract added) there was no apparent change—but at the end of this time the mixture lost its diffluent condition, and became converted into a gelatinous mass, exactly resembling a thin starch jelly. Bye and bye the

---

* In cooking this solution it is advisable to use the water-bath, for otherwise some of the albumen coagulates and cakes on the bottom of the vessel, and the liquid froths up in an inconvenient manner. If the eggs from which the albumen has been obtained are not freshly laid some coagulation will occur on boiling the solution. The addition of a drop of ammonia obviates this.

gelatinous matter broke up into little masses, which floated in a transparent liquid. At this point the action seemed to be arrested. The floating masses of jelly remained almost undiminished in quantity after twenty-four and even forty-eight hours. When the mixture was filtered the liquid portions came through perfectly transparent, and the jelly-like matter was left on the filter. The filtrate was found to be a rich and nearly pure solution of peptone—uncontaminated with any undigested or half-digested albumen. The jelly-like matter was found to be insoluble in water, hot or cold, but it dissolved readily in acids, and was rapidly digested by pepsin and hydrochloric acid.

If a large amount of pancreatic extract was used, a considerable proportion of the jelly-like matter was slowly dissolved—but the main result was always the same—the pancreatic ferment was only able to convert a part of the albumen into peptone, whereas the gastric ferment converted the entire quantity, with the exception of an insignificant residuum.

In the case of milk, the relation of the two ferments is reversed. Tryptic digestion of milk is rapid, and leaves only a very slight residue—whereas peptic digestion is slow, and leaves a large residue. I have some further observations to make on the digestion of milk by trypsin, but it will be more convenient to take up the subject when I come to the use of pancreatic extract for the preparation of artificially digested food.*

[The primary function of the pepsin-acid of gastric juice is evidently to get the albuminoid matters into solution rather than to peptonise them. Bernard ranked pepsin low as a peptonising agent. He looked on gastric digestion as a hasty preparatory process, introductory to the more perfect intestinal digestion. This seems to me a truthful view. The rapidity with which boiled egg-albumen is dissolved by an active preparation of pepsin is very striking—but the matter in solution is not, as is well known, really digested—it is merely liquefied and is nearly all re-precipitated by neutralization. The later stages of digestion—those which approach or reach up to peptone—appear to be performed by pepsin very slowly and as it were with difficulty. The case is quite otherwise with trypsin. The action of trypsin on solid albuminoid masses is exceedingly

---

* Most observers have noticed the occurrence of this indigestible residuum (named *Dyspeptone* by Meissner) in the artificial gastric digestion of proteids. I have noticed the same in the digestion of milk by pancreatic extract.

slow and imperfect—but its action on liquid casein, as it exists in milk, is marked by a rapidity and completeness of which there is no parallel example in the case of pepsin.]

Peptogens.

I may here advert to a singular view advanced by Schiff in regard to the production and secretion of pepsin and trypsin. Schiff found that when an insoluble aliment such as white-of-egg (or fibrin, or meat which had been deprived of its soluble portions) was introduced into the stomach of a fasting animal no pepsin was secreted, and the albumen remained undigested; but if with the albumen certain soluble aliments were introduced into the stomach, then pepsin was produced, and digestion immediately began.

To these substances, which had the power of provoking the formation and secretion of pepsin, Schiff gave the name of *peptogens*. Among the most effective peptogens were found to be solutions of dextrine, extract of meat (or soup), infusion of green peas, bread (which contains dextrine), gelatin, and peptones. On the other hand solutions of grape-sugar, soluble starch, fat-emulsion or gum had no peptogenic effect; and milk and coffee had not much. Schiff further found that peptogenic substances were just as effective when they were injected into the blood, or into the cellular tissue, or introduced as enemata into the rectum, as when they were introduced directly into the stomach. On the other hand when peptogens were injected into the small intestine their influence was not observed—their effect seemed to be annulled by some action of the mesentric glands, or by some change induced in them in their passage along the throracic duct.

On the ground of these experiments—and they were numerous and oft repeated, and gave constant and decisive results— he concluded that the absorption into the blood of these soluble aliments was a necessary preliminary of proteid digestion—that no pepsin or trypsin was secreted unless these substances existed beforehand in the blood. The first act, according to Schiff, of gastric digestion was the absorption from the constituents of a meal of these soluble peptogens by the veins of the stomach. On this followed immediately the secretion of pepsin and the commencement of digestion proper.*

---

* See *Leçons sur la Physiologie de la Digestion*, Paris, 1867, T. II., p. 200 et seq.— where the experimental evidence on which he relies is set forth at length.

These views and experiments of Schiff have not been allowed to pass without challenge, but they have not yet been overturned. If they should be substantiated they will give, curiously enough, a scientific sanction to the prevailing custom of commencing dinner with soup.

THE MILK-CURDLING FERMENT.

You all know that one of the most striking properties of gastric juice is to curdle milk. This property is utilized on a large scale in the industrial art of making cheese. Rennet, which has been used for that purpose from remote antiquity,* is simply an infusion of the fourth stomach of the calf in brine. The curdling of casein by rennet does not depend upon the acid of the gastric juice, for it takes place when the milk is neutral or even faintly alkaline. It has until lately been believed that this property was an inherent attribute of pepsin, but this opinion is no longer tenable. Brücke succeeded, by a process I need not particularize, in producing pepsin which had an energetic action on proteids, but which did not possess, except in the feeblest degree, the power of curdling milk. Mr. Benger also found that an extract of pig's stomach in saturated brine, while it possessed energetic action as a milk-curdler, had comparatively only feeble proteolytic powers. We must therefore regard the agent in gastric juice which curdles milk as a distinct substance from pepsin.

In the course of my experiments on pancreatic extract, I made the unexpected observation that the pancreas also contained an agent capable of curdling milk. I found this property in the pancreas of the pig, the sheep, the calf, the ox, and the fowl. In whatever way the extract of the gland was made, whatever solvent was used, this property of curdling milk was present in it; but the brine extract exceeded all others in curdling capacity. If a few drops of extract of pancreas be added to some warm milk in a test-tube the milk becomes a solid coagulum in a few minutes. Some minutes later the whey begins to separate from the curd. In short the action resembles exactly that of calf's rennet; and, so far as I know, you could make cheese with pancreatic rennet

---

* Cheese was in use among the ancient Hebrews. When Jesse sent David to visit his brethren in the camp of Saul, and to bring them corn and bread, he also instructed him to "carry these ten cheeses unto the captain of their thousand."—1 Sam., xvii. 18.

as perfectly as you can with gastric rennet. There is, however, not an absolute identity of the two agents. I said just now that gastric rennet produced curdling in neutral and even in faintly alkaline milk ; but if the alkali exceed a very small proportion, ordinary rennet does not curdle milk. I found that an alkalescence exceeding that produced by one grain of bicarbonate of soda to an ounce of milk altogether prevented the milk being curdled by gastric rennet. But this is not so with pancreatic rennet. You may add two, three, or four grains of bicarbonate of soda to each ounce of milk, and still the pancreatic rennet will induce curdling with undiminished energy. Milk is likewise curdled by pancreatic extract when quite neutral, and even when very faintly acid. Indeed, it appeared to me that a very faintly acid milk curdled more actively with pancreatic extract than neutral milk, but not so actively as alkaline milk.

That the curdling agent of the stomach and pancreas is a true ferment, and not some inorganic chemical agent, seems to be proved by the fact that boiling or even heating to 160°F. (70°C.) instantly destroys its power. I found moreover that like other soluble ferments it is precipitated, but not truly coagulated, by alcohol—for it recovers its solubility and activity when the alcohol is removed, even after a contact of several weeks.

The curdling ferment of the pancreas is a distinct body from trypsin, as the following experiments show. (1) Some brine extract of pancreas (which was known to possess strong proteolytic energy) was acidulated with hydrochloric acid in the proportion of 1 per 1000, and then placed in the warm chamber at a temperature of 104°F. (40°C.) for a period of three hours. It was then carefully neutralized with bicarbonate of soda. When thus treated, the extract was found to have lost its proteolytic power, but its curdling action on milk was almost as energetic as ever. (2) A portion of the same brine extract of pancreas was filtered under vacuum pressure through porous earthenware. The filtered product was found to possess an undiminished faculty of curdling milk, but it had almost no power of dissolving the curds. The curdling ferment had evidently traversed the earthenware freely, but only traces of trypsin had passed through.

What is the real function of the curdling ferment? Seeing its striking reaction with milk, one's first idea is that it must have something to do with the digestion of casein.   But a little consideration shows that this idea is altogether improbable. Although all mammalia start life on a milk diet, milk does not form a part of the normal diet of any adult creature except man.   Nor can its universal presence in the mammalian digestive organs be regarded as a vestigial phenomenon—a "memory" of the suckling phase of their existence—for the same curdling property is found in the stomach and pancreas of the fowl which never at any period of its life fed on milk. Moreover, it may be doubted whether the ferment in question is the actual agent which curdles milk on its passage into the stomach—for the acid of the gastric juice, which also curdles milk, would probably be beforehand with it, inasmuch as its action is a good deal more prompt than that of the ferment. In the pancreatic digestion of milk, the occurrence of curdling has appeared to me to be a distinct hindrance to the process. Has this ferment any true digestive functions?   I think this is quite open to doubt.   Its action on milk is apparently akin to that of the fibrin-ferment on blood—and it may likewise have some kindred purpose—but what that purpose may be I am unable to conjecture.

EMULSIVE FERMENT.—DIGESTION OF FATS.

The digestive change undergone by fatty matters in the small intestine consists mainly in their reduction into a state of emulsion, or division into infinitely minute particles.   In addition to this purely physical change, a small portion undergoes a chemical change whereby the glycerine and fatty acids are dissociated.   The fatty acids thus liberated then combine with the alkaline bases of the bile and pancreatic juice, and form soaps.   The main or principal change is undoubtedly an emulsifying process, and nearly all the fat taken up by the lacteals is simply in a state of emulsion, and not of saponification.   It is however quite certain that both these processes do take place in the small intestine, though in very unequal degrees.   The only question in connection with the digestion of fat which I propose to examine is :—Whether these changes are produced by the operation of a soluble ferment, or by some other and different agencies.   In his latest utterances on this subject, Bernard[*] insisted that the digestion of fat, like the digestion of

* Claude Bernard.   Leçons sur les Phénomènes de la vie.   T. ii., p. 346. Paris, 1879.

starch and proteids, consisted in the action of a soluble ferment, which he named *Ferment Emulsif*. This ferment he alleged, first emulsified and then saponified fats. In the intestine the change scarcely went beyond emulsion—in this condition fat was found in the contents of the lacteals. Saponification took place almost exclusively further on, and later, in the blood. It is certainly established that the pancreatic juice exercises a marked influence on the digestion of fats, and it is in the pancreas, according to Bernard, that the emulsive ferment is to be found. Bernard demonstrated that healthy pancreatic juice has quite a special faculty of emulsifying fats. Pancreatic tissue has also the same property. If a portion of fresh pancreas be rubbed up with fatty matter and water, you get an emulsion which is quite persistent. I have not had an opportunity of examining the behaviour of pancreatic juice with fatty matter, and cannot therefore speak of its properties; but it is singular if, as alleged, the effect of pancreatic juice and pancreatic tissue on fat be due to the presence of a soluble ferment, that the extracts of pancreas possess none of the same power. I have made extracts of pancreas in various ways—with simple water, with chloroform-water, with dilute spirit, with solutions of boracic acid, of borax, and of both combined, with glycerine and water, with brine, and with solution of salicylic acid, and of salicylate of soda; and yet I could not satisfy myself that any of these extracts possessed any special power of emulsifying fats, nor of liberating the fatty acids and inducing saponification. Paschutin* states that the emulsive ferment of the pancreas can be extracted by a solution of bicarbonate of soda. An extract of pancreas made by myself with a two per cent solution of bicarbonate of soda was indeed found to have a very marked emulsifying power, but it had the same power, even in an enhanced degree, after being boiled, which showed that its emulsifying properties could not depend on the presence of a soluble ferment.

I was equally unsuccessful in my attempts to verify the alleged power of extracts of pancreas, and of crushed pancreatic tissue, to liberate the fatty acids. When fresh pancreas, finely triturated with sand, was digested with milk in the warm chamber, I could not obtain satisfactory evidence of the development of free acid from the decomposition of the fat of the

---

* Hoppe-Seyler, Physiologische Chemie, p. 257. Berlin, 1878.

milk by a soluble ferment. The pancreas itself yields a slightly acid solution when infused in water, and a mixture of milk and pancreatic tissue always showed a slight acid reaction ; but when this primary acidity was neutralized, no further production of acid took place until such a time had elapsed as was sufficient to permit the development of organised ferments and the origination of the lactic fermentation. If the development of organised ferments was prevented by the addition of antiseptics—such as chloroform or a combination of boracic acid and borax—a mixture of milk and crushed pancreas remained neutral for several days. The same results followed when I operated on emulsions made with crushed pancreas and lard or olive oil. In my numerous observations on the digestion of milk with various pancreatic extracts, I never could detect the production of an acid reaction, unless organised ferments were permitted to intervene.

I obtained similar negative results with almond emulsion. Bernard attributed the formation of an emulsion when almonds (or other oily seeds) were rubbed up with water to the presence in the seeds of a soluble ferment. But I found, to my surprise, that almonds which had been boiled for seven hours still produced a perfect emulsion. As all known soluble ferments are destroyed by boiling, this result seems irreconcilable with Bernard's view. I also found that almond emulsion kept in the warm chamber for six or eight hours at a temperature of 100° F. (38° C.) showed not the slightest evidence of an increase of its original faintly acid reaction. It appeared to be more probable that the fatty matter in the almond existed in the seed in the condition of a solid emulsion, and that the formation of a fluid emulsion by trituration with water was due simply to the liberation of the minutely divided oil particles, rather than to the intervention of a soluble ferment.

It is with considerable hesitation that I venture to place myself even in apparent contradiction with so great an observer as was Claude Bernard ; and I by no means pretend that these observations traverse the main conclusions for which he contended as to the digestive transformations of fat in plants and animals. The views which Bernard developed on the digestive process are based on inductions so wide, and observations so multiplied, that I feel satisfied that their substantial accuracy will be ultimately established in regard to fat, as they have already been established in regard to starch and cane-sugar.

Some observations made by Brücke promise to throw a fresh light on the digestion of fat. Brücke found that oils and fats which contained an admixture of free fatty acids—in other words, which were more or less rancid—were emulsified by a slight agitation with a weak solution of carbonate of soda. J. Gad extended these observations, and showed that even simple contact of a rancid oil with the alkaline solution was sufficient to effect a mechanical division of the oily matter. I have repeated these observations, and the results are certainly remarkable. The different behaviour of two specimens of the same oil, one perfectly neutral and the other containing a little free fatty acid, is exceedingly striking. I have here before me two specimens of cod-liver oil—one of them is a fine and pure pale oil, such as is usually dispensed by the better class of chemists; the other is the brown oil sent out under the name of De Jongh. I put a few drops of each of these into these two beakers, and pour on them some of this solution, which contains two per cent of bicarbonate of soda. The pale oil, you see, is not in the least emulsified; it rises to . the top of the water in large clear globules; the brown oil, on the contrary, yields at once a milky emulsion. The pale oil is a neutral oil, and yields no acid to water when agitated with it—in other words, it is quite free from rancidity; but the brown oil, when treated in the same way, causes the water with which it is shaken to redden litmus paper. I was surprised to find that olive oil (salad oil), which appeared quite sweet, and had not the slightest taste or smell of rancidity, gave a milky emulsion with the soda solution. This oil did not yield any acid reaction to water when agitated therewith. Nevertheless it evidently contained a little free fatty acid (probably oleic acid, which is insoluble in water, and therefore does not acidify water shaken up with it), for when a portion of this oil was washed with a strong solution of carbonate of soda, and then allowed to separate, the oil thus freed from acid no longer gave an emulsion with the weak soda solution. It would appear that an admixture of only a very small proportion of free fatty acid is sufficient to induce emulsification—a quantity so small as not to cause any appreciable rancidity to the senses of smell or taste. This specimen of almond oil is to all appearance perfectly sweet, but it communicates a rather sharp acid reaction to water shaken up with it, and it gives, as you see, an instantaneous and perfect emulsion with the soda solution.

The bearing of these observations on the digestion of fat is plain. When the contents of the stomach pass the pylorus they encounter the bile and pancreatic juice, which are alkaline, from the presence in them of carbonate of soda. So that the fatty ingredients of the chyme, if they only contain a small admixture of free fatty acids, are at once placed in favourable circumstances for the production of an emulsion without the help of any soluble ferment, the mere agitation of the contents of the bowel by the peristaltic action being sufficient to effect the purpose.

This view of the matter renders it necessary that fresh inquiries should be made into the effect of gastric digestion on fats. It has hitherto been supposed that fatty and oily substances undergo no change in the stomach, but it is quite possible that something may have been overlooked. It was noted by Richet in the patient with a gastric fistula that the fatty matters were detained a long time in the stomach, and that they only passed through the pylorus with the last portions of the meal. It is also a familiar experience to most dyspeptics that rancid eructation are a frequent occurence in the later stages of gastric digestion. If it should turn out that among the complex operations taking place in the stomach there occured some slight decomposition of the neutral fats, and a liberation of a small quantity of free fatty acid, such a result would supply the necessary condition for the emulsification of the neutral fats in the duodenum. In speculating on this subject it is difficult to shut one's eye to the possibility of the intervention of formed or organised ferments in the digestive process. It is well known that fatty acids are liberated in the decomposition of neutral fats by bacteroid ferments (zymophytes), and the presence of ferments of this class in the living stomach has been so repeatedly observed that it may well give rise to the suspicion that they are a normal ingredient of the gastric mucus, and have a normal function to perform in the digestion of some portions of our food. It is not, however, desirable to push speculations of this kind in advance of observed facts, and I only mention them as hints for further inquiry in regard to the digestion of fat.

LECTURE III.

SUMMARY.—Effects of Cooking on Food—Artificially Digested Food—Pancreatic Digestion of Milk—Modified Casein or Meta-casein—Directions for the Preparation of Peptonised Articles of Food—Peptonised Milk—Peptonised Gruel—Peptonised Milk-Gruel—Peptonised Soups, Jellies, and Blanc-manges—Peptonised Beef-Tea—Nutritive Value of Peptonised Food—Clinical Experience of Peptonised Food—Uræmic Vomiting—Gastric Catarrh—Crises of Cardiac Disease—Pernicious Anæmia—Gastric Ulcer—Pyloric and Intestinal Obstruction—Use of Pancreatic Extract as an Addition to Food shortly before it is Eaten—Pancreatic Extract as an Addition to Nutritive Enemata.

The suggestion to administer to invalids artificially digested food appears at first sight a somewhat startling proposal. So great an interference with the order of nature would seem to go beyond the legitimate province of art. But when we reflect how largely art already interferes in the preparation of our food, the taking of this further step will appear less surprising. The practice of cooking is in reality as complete a departure from the ways of untutored nature as artificial digestion would be. Among the almost countless species of animals, not one of them, except man alone, cooks his food, insomuch that man has, not inaptly, been defined as the cooking animal.

Effects of Cooking.—The process of cooking fulfils far more important ends than that of improving the savour of food—far more important even than the mechanical disintegration which generally attends the process. It produces certain chemical changes in several of the most important alimentary principles which render them incomparably more susceptible to the action of the digestive ferments than in the uncooked state. The discovery of the use of fire-heat in the preparation of his food must indeed have constituted one of the earliest and most important steps in the process by which man has emerged from the ranks of the dumb creation. The stores of proteid and farinaceous nutriment contained in the seeds of cereals and leguminous plants, and in the bulbs, tubers, roots, and succulent stems of certain vegetables are, in the raw state, nearly altogether beyond his powers of digestion. By the discovery of the art of cooking these immeasurable stores were at one stroke laid open to him. It is moreover chiefly by the same art that he has been enabled to take his food at intervals, in separate meals, and has thereby been for ever relieved from the necessity which is imposed on all animals in the wild state of having to spend almost the entire of their waking hours either in seeking after their food, like the carnivora, or in consuming their food like the vegetable feeders. This immunity secured to him the untold advantage of possessing the leisure requisite for the cultivation of his higher faculties.

The practice of cooking is not equally necessary in regard to all articles of food. There are important differences in this respect, and it is interesting to note how correctly the experience of mankind has guided them in this matter. The articles of food which we still use in the uncooked state are comparatively few ; and it is not difficult in each case to indicate the reason of the exemption. Fruits, which we consume largely in the raw state, owe their dietetic value chiefly to the sugar which they contain ; but sugar is not altered by cooking. Salads may be regarded more as a relish for other food, and as having a *quasi* medicinal purpose, rather than as a substantial source of nutriment. Milk is consumed by us both cooked and uncooked, indifferently, and experiment justifies this indifference, for I found on trial that the digestion of milk by pancreatic extract was not appreciably hastened by previously boiling the milk.

Our practice in regard to the oyster is quite exceptional, and furnishes a striking example of the general correctness of the popular judgment on dietetic questions. The oyster is almost the only animal substance which we eat habitually, and by preference, in the raw or uncooked state ; and it is interesting to know that there is a sound physiological reason at the bottom of this preference. The fawn-coloured mass which constitutes the dainty of the oyster is its liver, and this is little else than a heap of glycogen or animal starch. Associated with the glycogen, but withheld from actual contact with it during life, is its appropriate digestive ferment—the hepatic diastase. The mere crushing of the dainty between the teeth brings these two bodies together, and the glycogen is at once digested without other help by its own diastase. The oyster in the uncooked state, or merely warmed, is, in fact, self-digestive. But the advantage of this provision is wholly lost by cooking, for the heat employed immediately destroys the associated ferment, and a cooked oyster has to be digested, like any other food, by the eater's own digestive powers.

With regard, however, to the staple articles of our food, the practice of cooking it beforehand is universal. In the case of farinaceous articles cooking is actually indispensable. When men under the stress of circumstances have been compelled to subsist on the uncooked grain of the cereals, they have soon fallen into a state of inanition and disease. By the process of cooking, the starch of the grain is not merely liberated from its protecting envelopes, but it suffers a chemical change, by which it is transformed into the gelatinous condition, and this immensely facilitates the attack of the diastasic ferments. A change of equal importance seems to be induced in the proteid matter of the grain. I found that the gluten of wheat was greatly more digestible, by both artificial gastric juice and by pancreatic extract, in the cooked than in the uncooked state. In regard to flesh meat the advantage of cooking consists chiefly in its effects on the connective tissue and the tendinous and aponeurotic structures associated with muscular fibre. These are not merely softened and disintegrated by cooking, but are chemically converted into the soluble and easily digested form of gelatin. I made some instructive observations on the effects of cooking on the contents of the egg. The change induced by cooking on egg albumen is very striking. For the purpose of testing this point I employed the solution of egg albumen before spoken of, made by mixing white-of-egg with nine times its volume of water. This solution when boiled in the water-bath does not coagulate nor sensibly change its appearance, but its behaviour with the digestive ferments is completely altered. In the raw state this solution is attacked very slowly by pepsin and acid, and pancreatic extract has almost no effect on it; but after being cooked in the water-bath, the albumen is rapidly and entirely digested by artifical gastric juice, and a moiety of it is rapidly digested by pancreatic extract.

My object in making these remarks is to show that the changes impressed on food by cooking form an integral part of the work of digestion—a part which we of the human race get done for us by the agency of fire-heat—but a part which the lower animals are compelled to perform by the labour of their own digestive organs. It must also be borne in mind that the digestive process carried on in the alimentary canal is, strictly speaking, executed on a doubling of the exterior surface, and not in the true interior of the body. If we take all these considerations into account, it will appear, I think, not unnatural

D

that we should try to help our invalids by administering their food in an already digested, or partially digested, condition. We should thereby only be adding one more to the numberless artificial contrivances with which our civilized life is surrounded.

Dr. Pavy* was, I believe, the first to carry into actual practice the idea of preparing an artificially digested food. At his suggestion, Messrs. Darby and Gosden introduced a preparation which consisted of meat reduced to a fluid state by artificial digestion. The formula for this preparation has not, so far as I know, been made public. It is still in the market, and is sent out and advertised by Savory and Moore under the title of "Darby's Peptonised Fluid Meat." A specimen of this preparation is on the table before me. It has the appearance of a light brown very thick treacle. It has a strong salt taste and an agreeable meaty flavour, without any bitterness. It is mostly soluble in water. The solution does not precipitate on boiling, nor with nitric acid, and it gives a strong reaction of peptone, with Fehling's solution, and with tannin.

I have also before me a specimen of artificially digested meat prepared by Mr. Benger. It is a greyish looking extract, with a pleasant meaty flavour, and is quite devoid of bitterness. It dissolves mostly in water, and the solution gives the usual reactions of peptone in great intensity. Mr. Benger informs me that it is made by operating, at a temperature of about 140° Fahr., on finely triturated raw beef with pancreatic extract and a little carbonate of soda. The solution thus obtained is neutralized with hydrochloric acid, and then evaporated at 212° F. to the consistency of a solid extract. Both of these preparations appear to me to be much superior in value to any of the meat extracts hitherto introduced.

But however useful preparations of this class may prove to be, in a limited range of circumstances, it is pretty evident that if artificially digested food is to be employed on the large scale, and among all classes, means must be found to bring the preparation of it within the range of culinary operations and the apparatus of the kitchen and sick-room.

The difficulty hitherto encountered in the production of an

* A Treatise on Food and Dietetics. 2nd ed., p. 559.

artificially digested or peptonised* food, suitable for invalids, is mainly owing to the use of the gastric method in its preparation. If you subject any native article of food—say milk, bread, egg, or meat—to artificial digestion with pepsin and hydrochloric acid, you destroy more or less completely the grateful odour and taste, and the inviting appearance, which made it desirable as food, and convert it into an unsavoury mess, from which the human palate turns away with disgust. The unsavouriness of artificially digested food is, however, not due to any ill taste or smell inherent in the products of digestion—which, when purified, are both odourless and free from any unpleasant flavour— but to a number of bye-products of various kinds, which accumulate as digestion proceeds. One of these bye-products is a substance with a pure bitter flavour, which seems to be a constant accompaniment of gastric digestion. It is also developed in some cases in the later stages of pancreatic digestion. I have not observed this bitter substance, except in the digestion of proteids. It is evidently a normal educt of the process, and its presence probably accounts, in most cases, for the bitter flavour of the eructations, of which dyspeptics complain, and which is generally attributed to the regurgitation of bile. It would be interesting to know more about this subject.

My own efforts to produce a palatable peptonised food have been chiefly directed to the pancreatic method. The pancreas excels the stomach as a digestive organ, in that it has power to digest the two great alimentary principles, starch and proteids ; and an extract of the gland is possessed of similar endowments. This double power is a manifest advantage in dealing with vegetable aliments, which contain both starch and proteids.

Any extract of pancreas may be used for the preparation of artificially digested food, but the most suitable are those prepared with dilute spirit or chloroform water. The extract sent out by Mr. Benger, under the name of "Liquor Pancreaticus,"† is an almost faultless pharmaceutical preparation. It is made by extracting perfectly fresh and finely chopped pancreas, with four times its weight of dilute spirit. By some ingenious devices, Mr. Benger has succeeded in overcoming the mechanical difficulties of the manufacture, and has produced an extract

---

*I may be permitted to use the word "peptonised" as a convenient abbreviation for the phrase "artificially digested."

† Sent out by Mottershead & Co., Chemists, Manchester.

which possesses the diastasic and proteolytic properties of the pancreas in a highly concentrated degree. It is a nearly colourless solution, with very little taste or smell beyond that of the spirit used to preserve it. It is of this preparation that I must be understood to speak in what I have now to say on the production of artificially digested food by pancreatic extracts.

My attention was first turned to the artificial digestion of milk, and I soon found that it was possible, by means of pancreatic extract, to digest this important article of food with comparatively little disturbance of its taste and appearance. Milk contains all the elements of a perfect food, adjusted in their due proportion for the nutrition of the body. Two out of its three organic constituents—namely, the sugar and the fat—exist already in the most favourable condition for absorption, and require little, if any, further assistance from the digestive ferments. It is therefore obvious that if we could change the casein of milk into peptone without materially altering the flavour and appearance of the milk, such a result would go far towards solving the problem of supplying an artificially digested food for the use of the sick.

Pancreatic Digestion of Milk.

When milk is subjected to the action of pancreatic extract, at a temperature of 100° F. (38° C.), in an open glass vessel, a series of changes take place in it, which are highly interesting to watch. The first thing that arrests the observer's notice is that the tough wrinkled skin which quickly forms on the surface of warm milk when exposed to the air is either not produced at all, or is only produced in a very imperfect degree. In its stead there forms a slight brittle and perfectly smooth pellicle of quite a different appearance. The next thing noticed is that the milk becomes more or less softly curdled. Bye and bye the curds begin to redissolve, and the milk gradually reassumes its originally diffluent condition. A portion of the curd is, however, very resistent, and remains undissolved for many hours. If the milk be diluted beforehand with one-third or one-fourth of its bulk of water, this curdling phase is altogether omitted, or is only observed as a slight and transient thickening. Next follows a very curious change of aspect. The milk loses its glossy white appearance, and gradually assumes a dull yellowish-grey shade which is characteristic, and the degree of which enables the practised eye to judge with considerable precision how far the peptonising process has advanced. This change of aspect is, however, by no means conspicuous, and would scarcely be remarked in a cursory observation, except by comparison with unaltered milk. While these changes are proceeding, the milk gradually loses its proper flavour, and at length developes a pure bitter taste, which is to many palates not disagreeable. No really unpleasant flavour is produced, unless the process is allowed to go on to incipient decomposition.

The progressive transformation of the casein into peptone, of which these outward signs are the indications and accompaniments, may be followed by testing the milk from time to time with acetic acid. At first the addition of the acid causes an abundant precipitation of curdy matter, but this reaction progressively diminishes in intensity until at length it ceases altogether. When this point is reached the transformation may be regarded as complete. All the casein has been changed into peptone—even nitric acid no longer causes a precipitate. The time occupied in the transformation (supposing the temperature and the activity of the preparation to be constant) depends on the quantity of pancreatic extract added, and may be made to vary from a few minutes to several hours. In the ordinary operation of preparing peptonised milk for invalids, two or three teaspoonfuls of the liquor pancreaticus are added to one pint of milk, diluted with a third of its bulk of water. With these proportions, and at a temperature of 100° Fahr., the process is usually completed in from two and half to three hours.

Modified Casein or Metacasein.—The conversion of casein into peptone does not take place by a direct transformation of the one body into the other. You all know that milk does not curdle or coagulate in the least degree on being boiled, but when milk is subjected to the action of pancreatic extract (provided no alkali is added), it speedily loses this negative property and curdles abundantly on boiling. This coagulation on boiling is most intense soon after the addition of the extract, and it very gradually diminishes in intensity as the action goes on, and ceases altogether about the same time that acetic acid ceases to cause a precipitate. It was moreover found that if the milk was boiled at the period of the greatest intensity of this reaction, and thrown on a filter, the whole of the albuminoid matter of the milk was caught on the filter in the form of curds, and the filtrate showed not the slightest reaction of casein. These reactions revealed the interesting fact that in the transformation of casein into peptone by pancreatic extract, the first step in the process is the conversion of casein into an intermediate body, and that it is this intermediate body which is subsequently gradually changed into peptone. This body may provisionally be called *metacasein*, signifying thereby that it is still casein, but in a modified condition. Metacasein is characterised by two reactions, which, taken together, serve to distinguish it from other proteid bodies—it is coagulated by boiling in neutral media, and it is precipitated in the cold by acetic acid. The conversion of casein into metacasein in pancreatic digestion takes place almost suddenly, as is shown by the following experiment:—5 cubic centimeters of pancreatic extract were added to 100 cubic centimeters of milk, diluted with one-fourth of its bulk of water, and maintained at blood-heat. The first slight, almost doubtful, evidence of coagulation on boiling, was perceived in three minutes; in five minutes coagulation on boiling was pronounced; and in nine minutes it had reached its maximum. From this point coagulation on boiling, and precipitation on the addition of acetic acid, diminished in intensity *pari passu* very gradually for a period of two hours, when both reactions finally ceased.

Taking these observations and reactions together, it is evident that the conversion of casein into metacasein constitutes a first and distinct step in the transformation of casein by trypsin, and that this step is antecedent to the further and slower changes by which metacasein is transmuted into peptone. It is impossible not to see in this a striking analogy with the sudden transformation of gelatinous starch into soluble starch under the action of diastase, as described in a previous lecture.

When milk is rendered slightly alkaline by the previous addition of a little bicarbonate of soda, no precipitation on boiling occurs during its digestion by pancreatic extract. But the metacasein is nevertheless produced, and its presence may be detected by carefully saturating the alkali, and then boiling. For although metacasein is precipitated on boiling when the solution is neutral, it is not precipitated when the solution is even slightly alkaline. This is the reason why, in preparing peptonised milk for the sick, it is desirable to add to it a small quantity of bicarbonate of soda.

The foregoing account of the behaviour of milk with pancreatic extract will greatly facilitate the comprehension of the practical rules which must be followed in preparing peptonised dishes for invalids.

DIRECTIONS FOR THE PREPARATION OF PEPTONISED ARTICLES OF FOOD.

Peptonised Milk.—A pint of milk is diluted with a quarter of a pint of water, and heated to a temperature of about 140° F. (60° C.). Two teaspoonfuls of liquor pancreaticus, together with twenty grains of bicarbonate of soda, are then mixed therewith. The mixture is then poured into a covered jug, and the jug is placed in a warm situation in order to keep up the heat. At the end of an hour or hour and half the product is raised to the boiling point. It can then be used like ordinary milk.

The object of diluting the milk is to prevent the curdling which would otherwise occur and greatly delay the peptonising process. The addition of bicarbonate of soda prevents coagulation during the final boiling, and also hastens the process. The purpose of the final boiling is to put a stop to the ferment action when this has reached the desired degree, and thereby to prevent certain ulterior changes which would render the product less palatable. The degree to which the peptonising change has advanced is best judged of by the development of the bitter flavour. The point aimed at is to carry the change so far that the bitter taste is distinctly perceived, but is not unpleasantly pronounced. As it is impossible to obtain pancreatic extract of absolutely constant strength, the directions as to time and quantity must be understood with a certain latitude.* The extent of the pep-

* *Note on the estimation of the activity of pancreatic extracts.*—The variable activity of preparations of the digestive ferments, even when freshly made, and the gradual deterioration which they all suffer on keeping, renders it very desirable that we should possess a means of quantitatively estimating their power. In regard to pepsin preparations we have long possessed a useful, though a somewhat rough and troublesome, method of estimating their value by the quantity of boiled white-of-egg which a given quantity is able to dissolve, with the aid of dilute hydrochloric acid, in a given time, and at a given temperature.

I have lately brought to completion a method of estimating the value of pancreatic extracts, of which a full account is given in a paper read before the Royal Society on the 5th May, 1881. I can do no more in this place tonising action can be regulated either by increasing or diminishing the dose of the liquor pancreaticus, or by increasing or diminishing the time during which it is allowed to operate. By skimming the milk beforehand, and restoring the cream after the final boiling, the product is rendered more palatable and more milk-like in appearance.

[*Preparation of peptonised milk in the cold.*—The action of pancreatic extract on milk goes on at the ordinary temperature of the air exactly in the same way as at blood-heat—except that

than give an outline of the method. As the pancreas contains two wholly distinct enzymes, or ferments—namely, pancreatic diastase and trypsin—two distinct estimates have to be made by two distinct proceedings.

The diastasic activity of pancreatic extracts is estimated by the quantity of a standard starch-mucilage, containing one per cent of pure potato starch, which can be changed, or transformed, to the point at which it ceases to give a colour-reaction with iodine, by one cubic centimeter of the extract to be tested, in a period of five minutes, at a temperature of 40° C. The diastasic value comes out as the number of cubic centimeters of the standard mucilage thus changed. The mean diastasic value of Benger's *liquor pancreaticus*, prepared on the large scale from pig's pancreas, was found to be 125. Extracts prepared with the same proportions of the pancreas of the ox and sheep gave a mean diastasic value, respectively, of 11 and 12. Human saliva gave a diastasic value varying from 10 to 17.

The tryptic activity of pancreatic extracts is estimated by a method based on the property acquired by milk undergoing pancreatic digestion, of curdling when boiled. This reaction is described on p. 57 of the present volume, and it may be designated as the *metacasein reaction*. The advent of this reaction occurs earlier or later according to the mutual relations of the four following factors, namely, the quantity of pancreatic extract employed, its activity, the quantity of milk operated on, and the temperature at which the process is conducted. The tryptic value of a sample of pancreatic extract is estimated as the number of cubic centimeters of milk, which are transformed to the onset-point of the metacasein reaction, in a period of five minutes, by one cubic centimeter of the extract, at the temperature of 40° C. The tryptic value of Benger's *liquor pancreaticus*, reckoned in this way, was found to oscillate between 50 and 70. Extracts prepared from the pancreas of the ox and sheep gave about the same tryptic value.

A rough estimate of the tryptic activity of a sample of pancreatic extract may be obtained by the following proceeding. Three fluid ounces of fresh milk are diluted with an equal quantity of water and brought to a temperature of 60° to 65° Fahr. One fluid drachm of the extract is then added (with stirring) to the diluted milk, and the time noted. At intervals of one or two minutes a portion of the mixture is boiled in a test-tube. If the extract possess the mean activity of Benger's *liquor pancreaticus*, curdling on boiling will occur in four or five minutes from the commencement of the experiment. If it be more active than this, curdling will occur in two or three minutes. On the other hand, if it be feebler than this, curdling will not occur until after the lapse of ten, twenty, or forty minutes. In these latter cases the experiment should be repeated, using two, three, or four drachms of the extract to the same quantity of milk, so as to bring the onset-point of the metacasein reaction within the compass of three to ten minutes. In this way a fair idea of the activity of the preparation may be gathered.

it is slower, and requires a longer time for completion. The cold method has, however, a convenience and simplicity which recommend it for general use in the sick-room. I have accordingly drawn up the following directions for the preparation of peptonised milk at a temperature of 60° to 65° Fahr., which may be regarded as the ordinary degree of warmth maintained in rooms occupied by invalids. In the winter season it will be necessary to slightly warm the ingredients beforehand in order to bring them to the due temperature, but in the warmer seasons the operations can be carried on without any preliminary heating.

A pint of milk is diluted with half a pint of *lime-water**—or with half a pint of water containing twenty grains of bicarbonate of soda in solution. To this are added three tea-spoonfuls of liquor pancreaticus. The mixture is then set aside in a jug or other convenient vessel, in the sick-room, for a period of three or four hours. At the expiration of this time the milk is far advanced in the process of digestion, and has developed a slightly bitter taste. It is now ready for use. It may be used cold, either alone, or with soda-water, which covers the bitterish taste remarkably well—or it may be warmed and sweetened for administration to infants.

If milk, thus prepared, is consumed at the period indicated—that is to say at the end of three or four hours, it need not undergo any final boiling—it is better indeed to use it without boiling, because the half finished process of digestion will still go on for a time in the stomach. But if milk thus prepared has to be kept much longer, it is advisable to raise it for a moment to the boiling point, so as to bring the action of the ferment to a termination and thus to prevent those ulterior changes which render the product disagreeable to the palate.

The process can be regulated with the utmost nicety by occasionally tasting the mixture, and watching the development of the bitter flavour—and it can be permanently arrested at any moment by heating the product to the boiling point.]

Peptonised Gruel.—Gruel may be prepared from any of the numerous farinaceous articles which are in common use—wheaten flour, oatmeal, arrowroot, sago, pearl barley, pea or lentil flour. The gruel should be very well boiled, and made

---

* I owe the suggestion to use lime-water to Dr. Watkins, of Newton-le-Willows.

thick and strong.   It is then poured into a covered jug, and
allowed to cool to a lukewarm temperature.   Liquor pancreaticus is then added in the proportion of two teaspoonfuls
to the pint of gruel.   At the end of three hours the product
is boiled, and strained.   The action of pancreatic extract on
gruel is two-fold—the starch of the meal is converted into
sugar, and the albuminoid matters are peptonised.   The conversion of the starch causes the gruel, however thick it may
have been at starting, to become quite thin and watery.   The
bitter flavour does not appear to be developed in the pancreatic digestion of vegetable proteids, and peptonised gruels
are quite devoid of any unpleasant taste.   It is difficult to
say to what extent the proteids are peptonised in the process
of digestion by pancreatic extract.   The product, when filtered,
gives an abundant reaction of peptone ; but there is a considerable amount left of undissolved material.   Most of this,
no doubt, consists of insoluble vegetable tissue, but it also
contains some unliberated amylaceous and albuminous matter.
Peptonised gruel is not generally, by itself, an acceptable food
for invalids, but in conjunction with peptonised milk (peptonised milk-gruel), or as a basis for peptonised soups, jellies, and
blanc-manges it is likely to prove valuable.

   Peptonised Milk-Gruel.—This is the preparation of which I
have had the most experience in the treatment of the sick, and
with which I have obtained the most satisfactory results.   It
may be regarded as an artificially digested bread-and-milk, and
as forming by itself a complete and highly nutritious food for
weak digestion.   It is very readily made.   First, a good thick
gruel is prepared from any of the farinaceous articles above
mentioned.   The gruel, while still hot, is added to an equal
quantity of cold milk.   The mixture will have a temperature
of about 125° F. (52° C.).   To each pint (550 cc.) of this
mixture, two teaspoonfuls of liquor pancreaticus, and twenty
grains of bicarbonate of soda are added.   It is then set aside
in a warm place for two or three hours, and finally raised to
the boiling point, and strained.   The bitterness of the digested
milk is almost completely covered in the peptonised milk-
gruel; and invalids take this compound, if not with relish,
without the least objection.

   [Since the first publication of these lectures, peptonised milk-
gruel has found favour with many practitioners, and has come

into considerable use among their patients. I find, however, that some persons fail to peptonise milk-gruel so as to make it acceptable to the palate and stomach of the invalid. This is entirely due to allowing the peptonizing process to go on too far. Artificial digestion, like cooking, must be regulated as to its degree ; and it is easy to regulate the degree of artificial digestion by the length of time during which the process is allowed to go on. It must be remembered that liquor pancreaticus (and every other form of pancreatic extract) is more or less variable in its activity, just as the fires used in cooking vary in their intensity, and that allowance must be made for this variability. If the liquor pancreaticus is very active the slight bitterness, whereby it is known that the process has been carried far enough, is developed in an hour, or less ; but if the preparation is not so active two or three hours may be required to reach the same point. It must further be borne in mind that the warmer the temperature at which the process is carried on the quicker is the action of the ferment. The practical rule for guidance is to allow the process to go on until a perceptible bitterness is developed, and not longer. As soon as this point is reached the milk-gruel should be raised to the boiling point, so as to put a stop to further changes.]

Peptonised Soups, Jellies, and Blanc-manges.—I have sought to give variety to peptonised dishes by preparing soups, jellies, and blanc-manges containing peptonised aliments. In this endeavour I have been assisted by a member of my family, who has succeeded beyond my expectations. She has been able to place on my table soups, jellies, and blanc-manges containing a large amount of digested starch and digested proteids, possessing excellent flavour, and which the most delicate palate could not accuse of having been tampered with. Soups were prepared in two ways. The first way was to add what cooks call "stock" to an equal quantity of peptonised gruel, or peptonised milk-gruel. A second and better way was to use peptonised gruel, which is quite thin and watery, instead of simple water, for the purpose of extracting shins of beef and other materials employed for the preparation of soup. Jellies were prepared simply by adding the due quantity of gelatin or isinglass to hot peptonised gruel, and flavouring the mixture according to taste. Blanc-manges were made by treating peptonised milk in the same way, and then adding cream. In preparing all these dishes it is absolutely necessary to complete the operation of peptonising the gruel or the milk even to the final boiling before adding the stiffening ingredient. For if pancreatic extract be allowed to act on the gelatin, the gelatin itself undergoes a process of digestion, and its power of setting on cooling is utterly abolished.

Peptonised Beef-tea.—Half a pound of finely minced lean beef is mixed with a pint of water and 20 grains of bicarbonate of soda. This is simmered for two hours. When it has cooled down to a lukewarm temperature, a tablespoonful of the liquor pancreaticus is added. The mixture is then set aside for three hours, and occasionally stirred. At the end of this time the liquid portions are decanted and boiled for a few seconds. Beef-tea prepared in this way is rich in peptone. It contains about 4·5 per cent of organic residue, of which more than three-fourths consist of peptone—so that its nutritive value in regard to nitrogenised materials is about equivalent to that of milk. When seasoned with salt it is scarcely distinguishable in taste from ordinary beef-tea.

The extreme solubility of digested products—whether of starch or of proteids—detracts from their acceptability to healthy persons. To them they appear thin and watery—they miss the sense of substance and solidity which is characteristic of their ordinary food. But to the weak invalid without appetite this sense of substance or thickening is generally an objection, and they take with more ease an aliment which they can drink like water. The jellies and blanc-manges, on the other hand, give to invalids of more power that sense of resistance and solidity which is desired by those of stronger appetite.

NUTRITIVE VALUE OF PEPTONISED FOOD.

At the outset of this inquiry we are met with the question : Is it certain that the ultimate products of digestion are of equal nutritive value to the mixed transitional products which are produced in succession, and probably absorbed as the food is gradually transformed in the alimentary canal?—in other words, are maltose and peptones alone as valuable to the economy as a mixture of these substances with the several dextrines and hemipeptones which are presented to the absorbent surfaces in the course of natural digestion?

With regard to the products of starch-digestion, no direct

experiments have been made on the nutritive value of maltose, and we can do little more than conjecture that the intermediate dextrines have a usefulness of their own. That they are absorbed seems proved (as might have been expected from their known diffusibility), for they have been detected in the blood, and especially in the blood of the portal vein.

With regard to peptones, we have more information. It was naturally assumed by the earlier observers who identified peptone as the chief ultimate product of digestion that this was the form under which proteids were taken up by the absorbents, and introduced into the blood for the nutrition of the tissues. But some ten years ago doubts were cast on this conclusion. It was alleged by Brücke and Voit that the nutrition of the tissues was maintained not by peptone, which was unfitted for this purpose, but by soluble albumen, which was absorbed in the undigested state from the primæ viæ, and that the office of peptone was a subordinate one, resembling that of gelatin, and consisted in aiding to preserve the tissue-albumen from too rapid destruction. Direct observations on the nutritive value of peptones have, however, shown this paradoxical view to be untenable. As the point is one of importance, I will endeavour to lay before you the proofs which have been already adduced of the food-value of peptones, and I will supplement these by some observations made by myself.

P. Plosz* was the first to put this question to the test of direct experiment. He fed a puppy dog weighing 1302 grams with an artificial compound made of fat and sugar, in imitation of milk, but in which the casein was replaced by artificially digested fibrin. In the course of eighteen days of this diet the dog grew and increased 501 grams in weight.

R. Maly† performed a similar experiment on a pigeon. He first fed the pigeon for several days on a regulated quantity of wheat until he had ascertained the quantity requisite to keep the bird in a state of nutritive equilibrium, in which it neither gained nor lost weight. He then made an artificial corn from starch, fat, gum, salts, and water, but in which the gluten was replaced by fibrin-peptone, and fed the pigeon on the same quantity of this artificial corn as he had before given of the natural corn. Under this novel diet the bird, after a short apprenticeship, not only maintained its weight, but actually put

* Pflüger's Archiv. f. d. ges. Physiologie : 1874, p. 323.

† Ibid, p. 585.

on flesh. This experiment seemed to show that peptone was even superior to natural gluten as a nutriment.

It was, however, objected to these experiments that they did not rigorously prove that peptone could build up the tissues, inasmuch as the increase of weight might have been due to an excessive accumulation of fat or of water, and that there might have been not an increase, but a decrease of the structural elements which contained nitrogen.

To meet these objections, Plosz and Gyergyai* instituted a third set of experiments on a dog weighing 2753 grams. The dog was brought down by a diet of simple water to a weight of 2531 grams. He was then fed on a mixture composed of sugar, starch, and fat, and containing, in addition, five per cent of purified peptone. Of this mixture about 400 grams were daily administered to the animal for a period of six days. In these six days the dog took in with his food 14·45 grams of nitrogen, but the total of the nitrogen excreted by the urine and fæces only amounted to 13·46 grams, so that nearly one gram of nitrogen had been retained in the body of the animal, which had increased in weight 259 grams. This experiment went to prove that peptone served to repair the wear and tear of the nitrogenised structural elements, and even contributed something to the increase of weight.

Adamkiewicz,† by a still more rigorous method, in a laborious series of experiments on a dog, which was fed on a diet wherein the only possible source of nitrogen was peptone prepared from blood-fibrin, arrived at the conclusion that peptone supplied nitrogen to the solid tissues, and that it possessed a nutritive value equal to albumen, or even slightly superior; and that therefore it could not be looked on as a bye product of digestion, but as the chief resultant of the transformation of proteids in the alimentary canal.

The experiments above cited seem sufficient to settle definitely the question raised by Brücke and Voit, and to establish the nutritive value of peptone as a source of nitrogen to the tissues. It seemed, however, desirable to obtain direct proof of the nutritive value of peptonised milk as compared with the natural article. My observations led me to the conclusion that it was easier to get a supply of peptonised aliment suitable for invalids by the artificial digestion of milk by pancreatic extract

* Pflüger's Archiv. Bd. x. 1875, p. 536.

† Natur und Nahrwerth des Peptons. Berlin, 1877.

than by any other method. But in so important a matter as the feeding of invalids, inference and conjecture seemed hardly a sufficient basis for actual practice.

The questions I proposed to myself were—(1) Is artificially digested, or peptonised, milk alone sufficient to sustain nutrition? and (2) is it as efficient in this respect as natural milk? For the purpose of answering these questions I procured four kittens, of the same brood, eight weeks old. Kittens at this age thrive perfectly on an exclusively milk diet. Two of the kittens were fed on natural milk, and the two others on milk previously digested by pancreatic extract. The digestion of the milk in the latter case was carried out to full completion—that is to say, until the milk became greyish and bitter, and no longer gave any precipitate with acetic acid, nor even with nitric acid. The animals were permitted to have as much of their respective foods as they could consume. The experiment was continued for a period of twenty days. The quantity consumed by each pair was as nearly as possible the same. All four continued in perfect health, and took their nutriment greedily. I was surprised to find that the pair fed on peptonised milk showed no repugnance to the bitter taste of their food, but appeared to relish it quite as well as their companions did their natural milk. The following table exhibits the gain in weight at the end of twenty days by each of the four kittens. The weights are given in grams, and in round numbers.

DIET.	Initial Weight.	Weight at the end of 20 days.	Increase of Weight	Increase of Weight per cent.
Natural Milk ... { No. 1—338	514	176	52
„ 2—558	835	280	50
Fully Peptonised { „ 3—403	626	223	55
Milk .... { „ 4—374	562	188	50

The Table shows that the percentage increase of weight was very nearly alike in all four animals. No difference could be perceived in the sleekness of their coats nor in the vivacity of

their gambols.   The experiment proves that peptonising the
milk in no degree spoils its nutritive value.   It also shows, as
might have been expected, that in healthy animals with abundance of digestive power there is no advantage to be gained from
the administration of food in a pre-digested state.   The quantity
of food which is taken and absorbed appears to be regulated not
so much by the quantity which can be digested as by the
quantity which can be assimilated.   This experiment had an
interesting sequel.   After Nos. 3 and 4 had been fed on *fully*
peptonised milk for a period of twenty days, they were fed for
a period of ten days on *partially* peptonised milk.   By using a
less quantity of the pancreatic extract in proportion to the milk,
and by allowing the process of artificial digestion to go on for
not more than half the time allowed in the previous experiment,
only a partial conversion of the casein was effected.   As nearly
as I could judge by the precipitate caused with acetic acid, the
casein was digested to about the extent of one half.   Under
these circumstances I certainly expected that Nos. 3 and 4
would gain weight as before, and would continue to thrive as
as well as Nos. 1 and 2.   It might have been anticipated that
under these new conditions the proper digestive powers of Nos.
3 and 4 would come into play and complete with ease the unfinished work of the artificial process.   But events did not pass
exactly as I anticipated.   The annexed table shows the actual
results.   The weights, as before, are given in grams.

DIET.	Initial Weight.	Weight on 3rd day.	Weight on 5th day.	Weight on 7th day.	Weight on 10th day.
	No.				
Natural	1—514	551—gain 37	570—gain 19	593—gain 23	618—gain 25
Milk.	2—838	900— ,, 62	936— ,, 36	993— ,, 57	1063— ,, 70
Half-pep-	3—626	693— ,, 67	687—loss 6	681—loss 6	693— ,, 12
tonised Milk.	4—562	645— ,, 83	638— ,, 7	648—gain 10	683— ,, 35

No interval was allowed to elapse between the two sets of
experiments, so that the initial weights in the second set are
the same as the terminal weights in the first set.   The abrupt

alteration in the diet of Nos. 3 and 4 produced a marked disturbance in their nutrition. But there was no declension in their general well-being, and they continued to take their modified diet with the same avidity as before. In the first three days Nos. 3 and 4 showed an unusual increase of weight—an increase quite out of proportion to that of their companions. Then came a check ; in the next four days they not only showed no gain, but showed even a slight loss of weight. In the last three days they began to gain weight again, and were evidently recovering from the temporary check caused by their change of diet. The explanation of this series of events which has presented itself to my mind is the following, but I by no means wish to lay stress on its accuracy. It seemed as if the enforced rest imposed on the digestive organs for a period of twenty days, during the use of fully pre-digested food, had temporarily diminished their natural vigour—just in the same way as the disuse of a limb diminishes for a time the strength of its muscles. During this first period, immediately following the change of diet, the unemployed digestive powers were suddenly called upon to resume their wonted activity ; but, owing to their enfeeblement from disuse, they were unable to respond to the call with due promptness, and undigested materials accumulated in the intestines. This accumulation accounted for the unusual increase in weight during the first period. During the next period the digestive embarassment made itself felt, and gave a check to the processes of growth and nutrition, and the animals ceased to grow and gain weight. During the last period the temporary enfeeblement of the digestive organs was passing away, in proportion as renewed exercise was restoring them to their original vigour, and growth and increase of weight again commenced.

Whether this explanation be trustworthy or not, the lesson indicated by it is probably a true one—namely, that except in extreme cases, when the digestive power is wholly lost or in complete abeyance, it is more advantageous to use a food which has been subjected to partial artificial digestion than food in which the process has been carried out to completion. If the patient possess any digestive power at all, it is better that that power should be kept in exercise than that it should be permitted to deteriorate still further from total disuse. This is in agreement with the rule we apply in other cases of en-

E

feebled function, according to which we endeavour to combine partial rest with moderate exercise.

Clinical Experience of Peptonised Food.

The extreme difficulty of arriving at reliable conclusions in regard to the effect of therapeutical agents is well known to every sober-minded inquirer. The difficulty is not less, but probably greater, in judging of the effects of dietetic means. I have now had a considerable experience, extending over a period of two years, of the use by invalids of peptonised milk and peptonised milk-gruel, and it would be an easy task for me to recite a string of cases in which improvement and recovery followed the use of these articles of diet. But evidence of this sort would be wholly delusive unless checked by such an analysis of the circumstances of each case as would enable me to isolate the influence of the dietetic means. In the immense majority of cases such an analysis would be obviously impossible, the conditions to be taken into account would be too numerous, and their relative influence too difficult to determine. In a question of this kind one is obliged, in a large degree, to fall back on general impressions, and on deductions based on physiological considerations. Nevertheless there are certain cases—most of them cases of incurable disease—in which the conditions are sufficiently simple to permit direct trustworthy conclusions to be drawn.

I found that peptonised milk-gruel was generally preferred, as being more agreeable to the palate, to simple peptonised milk; and by far the larger number of my observations were made with the former preparation. I was also soon satisfied that, with most rare exceptions, peptonised milk-gruel was perfectly acceptable to the invalid's stomach, and that a diet composed exclusively of this article could be used for many consecutive weeks without the slightest sign of failure of nutrition.

The cases in which the use of peptonised aliment appeared to produce the most striking benefits were those in which complete anorexia prevailed, and those in which the stomach was intolerant of food, and immediately rejected every form of nutriment. A brief review of the results obtained in cases of this kind will, I think, prove instructive.

Uræmic Vomiting.—In advanced Bright's disease incessant

vomiting is sometimes a distressing and intractable symptom. In some cases of this class I have seen the vomiting at once and permanently allayed by the use of peptonised milk-gruel. The downward course of the disease may not have been a moment checked, but the relief to the dying patient was great.

Gastric Catarrh.—That form of gastric catarrh which is the Nemesis of alcoholic excess often yields immediately to the use of peptonised food. In the later periods of cirrhosis there frequently prevails severe intolerance of every kind of food—the stomach rejecting even beef-tea snd diluted milk in the smallest quantities. The relief afforded by the use of peptonised milk-gruel in some of these cases is most striking—the vomiting ceases almost from the first, and the intolerable sense of distension diminishes.

Crises of Cardiac Disease.—Persons suffering from dilatation and valvular incompetancy usually encounter one or more crises which are susceptible of relief before finally succumbing to their disorder. These crises are marked by a general venous stagnation. with severe congestion of the lungs, liver, and kidneys, and rapidly rising dropsy. Associated with these symptoms, there is generally almost complete inability to take food, and sleeplessness. In this condition I have seen marked relief follow the use of peptonised aliment. I have long observed, as I doubt not have many of you, that the condition here described is often alleviated in the most striking manner by the use of exclusively liquid nourishment—such as milk or milk-gruel, given in small portions sub-continuously, or sippingly, as it were, throughout the waking hours—the patient being never permitted to take a distinct meal, nor a particle of solid food. As my practice has been to direct, in cases of this class, the administration of the peptonised aliment in this sipping fashion, the gratifying results noted have been partly due to the mode of administration ; but I have been convinced by more than one example, when the same liquid nourishment, in the natural and in the pre-digested condition, has been used in succession, that there was a distinct superiority in the pre-digested article.

Pernicious Anœmia.—In the earlier periods of this singular disorder, I am inclined to hope that pre-digested aliment may prove a valuable resource. In cases where the ailment, although fully declared, was still of comparatively recent origin, I have, in the last eighteen months, seen the disorder checked under the use of peptonised milk-gruel. In one case, owing to the

irritability of the stomach, the milk-gruel was at first administered per rectum with pancreatic extract, but was afterwards
tolerated by the stomach.   In three of these cases the ameliora-
tion went on to complete restoration.   In cases of longer stand-
ing I have failed by the same means to obtain the slightest improvement.

Gastric Ulcer.—The use of an exclusively liquid nourishment
given sub-continuously, in the manner before indicated, is a
well-known and most efficacious mode of treatment in these
cases.   But since adopting the plan of giving peptonised milk-
gruel, I think I have perceived that the results were distinctly
better than before, especially in cases associated with epigastric pain.   The almost absolute rest procured by this food
for the ailing organ appeared to be an additional advantage.   I
may be permitted to mention one case.   The patient had
suffered from copious and repeated hæmatemesis, and from
severe epigastric pain.   The irritability of the stomach was such
that the simplest nourishment, given in the smallest quantities,
was immediately rejected.   Peptonised milk-gruel was, however,
tolerated at once ; vomiting only occurred two or three times
during the two first days of the treatment, and then ceased, as
did likewise the epigastric pain.   This patient used no other
food for a period of six weeks, and took daily from two to three
quarts—with steady recovery of flesh and strength.

Pyloric and Intestinal Obstruction.—Peptonised aliment would
appear to be especially suitable for use in these cases, but, so
far, I have been somewhat disappointed in the results.   The
vomiting has generally been effectually controlled, but I have
not been able to convince myself, in cases of pyloric stricture,
that the fatal event was delayed even a single day.   When the
obstruction has been temporary, and due to a removable cause,
the results have been of course more satisfactory.

I should be glad to see a further trial made of peptonised, or
partially peptonised, milk in the gastric and intestinal catarrh
of infants.   In one severe case of this class a favourable result
was immediately obtained ; in another case there was greater
tolerance of food, and more comfort after it, than with the use
of simply diluted milk.   It would be interesting also to have
experience of the use of peptonised aliment in typhoid fever,
and in old age.   The greater variety which can now be given
to this form of food by the preparation of peptonised soups,
jellies, and blanc-manges will obviate the monotony sometimes

complained of under the continuous use of peptonised milk-gruel.

THE USE OF PANCREATIC EXTRACT AS AN ADDITION TO FOOD SHORTLY BEFORE IT IS EATEN.

The administration of pancreatic extract with, or immediately after, a meal can, I think, have only a limited utility. On entering the stomach the pancreatic ferments encounter the acid of the gastric juice, and when this rises above a certain point the activity of the ferments is destroyed. Still a not inconsiderable interval of time must elapse before this point is reached, and during this interval the pancreatic ferments can accomplish a certain amount of work. I have repeatedly administered pancreatic extract in this way, but I am unable to say positively that I have seen benefit from this mode of administration. There is, however, a modification of this plan, which I have lately put in practice, that promises better results. It is to add the extract to the food fifteen or twenty minutes before it is eaten. Certain dishes commonly used by invalids—farinaceous gruels, milk, bread-and-milk, milk flavoured with tea or coffee or cocoa, and soups strengthened with farinaceous matters, or with milk—are suitable for this mode of treatment. A teaspoonful or two of the liquor pancreaticus should be stirred up with the warm food as soon as it comes to table. And such is the activity of the preparation that even as the invalid is engaged in eating—if he eat leisurely as an invalid should do—a change comes over the contents of the cup or basin—the gruel becomes thinner, the milk alters a shade in colour, or perhaps curdles softly, and the pieces of bread soften. The transformation thus begun goes on for a time in the stomach, and one may believe that before the gastric acid puts a stop to the process the work of digestion is already far advanced.

This mode of administering pancreatic preparations is simple and convenient. No addition of alkali is required, and, of course, no final boiling. The only precaution to be observed is that the temperature of the food, when the extract is added, shall not exceed 150° F. (65° C.). This point is very easily ascertained, for no liquid can be tolerated in the mouth, even when taken in sips, which has a temperature above 140° F. (60° C.). If therefore the food is sufficiently cool to be borne in the mouth, the extract may be added to it without any risk of injuring the activity of the ferments.

Pancreatic Extract as an Addition to Nutritive Enemata.— Pancreatic extract is peculiarly adapted for administration with nutritive enemata. The enema may be prepared in the usual way with milk-gruel and beef tea, and a dessert-spoonful of liquor pancreaticus should be added to it just before administration. In the warm temperature of the bowel the ferments find a favourable medium for their action on the nutritive materials with which they are mixed, and there is no acid secretion to interfere with the completion of the digestive process.

I have now had some experience in this method of alimentation, and have been satisfied with its success. In one case a patient suffering from postpharyngeal abscess, which entirely occluded the œsophagus, was nourished exclusively for a period of three weeks (until the abscess broke) on enemata of milk-gruel mixed with pancreatic extract.

J. E. CORNISH, 33 PICCADILLY MANCHESTER.

J. E. Cornish's Publications.

ADAMSON (Prof. Robt., M.A.).
Roger Bacon: The Philosophy
of Science in the Middle Ages. 1*s.*

BRADLEY (S. Messenger, F.R.C.S.).
Injuries and Diseases of the
Lymphatic System. 5*s.*

ENGLAND (Edwin B., M.A.).
Greek Exercises for Beginners.
Translated from the Greek Grammar
of Prof. George Curtius. 1*s.*

HOPKINSON (Prof. Alfred, B.C.L.,
M.A.).
The Faculty of Laws, and the
Idea of Law. 6*d.*

LUND (Prof. Edwd., F.R.C.S.).
On Removal of the Entire Tongue,
by the Walter Whitehead Method,
with full details of the operation and
after treatment. 2*s.*

LUND (Prof. Edwd., F.R.C.S.).
Five Years' Surgical Work in
the Manchester Royal Infirmary. 3*s.*

LUND (Prof. Edwd., F.R.C.S.).
Suggestions for a Ready Method
of Recording Surgical Cases in Hospital Practice. With Plate. 1*s.*

MORGAN (Prof. John Edw., M.D.,
M.A. Oxon., F.R.C.P.).
The Victoria University: Why
are there no Medical Degrees? An
Address delivered to the Members of
the Owens College Medical Students'
Debating Society. 1*s.*

RANSOME (Arthur, M.D., M.A.).
The Present Position of State
Medicine in England. 1*s.*

REYNOLDS (Prof. Osborne, M.A.,
F.R.S).
Engineering Syllabus of the
Lectures at the Owens College; together with a Series of Examples
relating to the various subjects included in the Course. Arranged by
Mr. J. B. Millar, Assistant Lecturer
in Engineering. 3*s.* 6*d.*

ROBERTS (Prof. Wm., M.D., F.R.S.).
On Spontaneous Generation and
the Doctrine of Contagium Vivum. 2*s.*

WARD (Prof. A. W., LL.D.).
On Some Academical Experiences
on the German Renascence. 1*s.*

SCHORLEMMER (Prof. C., F.R.S.).
The Rise and Development of
Organic Chemistry, 2*s.* 6*d.*

Victoria University Calendar. 3*s.*